A MATERIAL LIFE

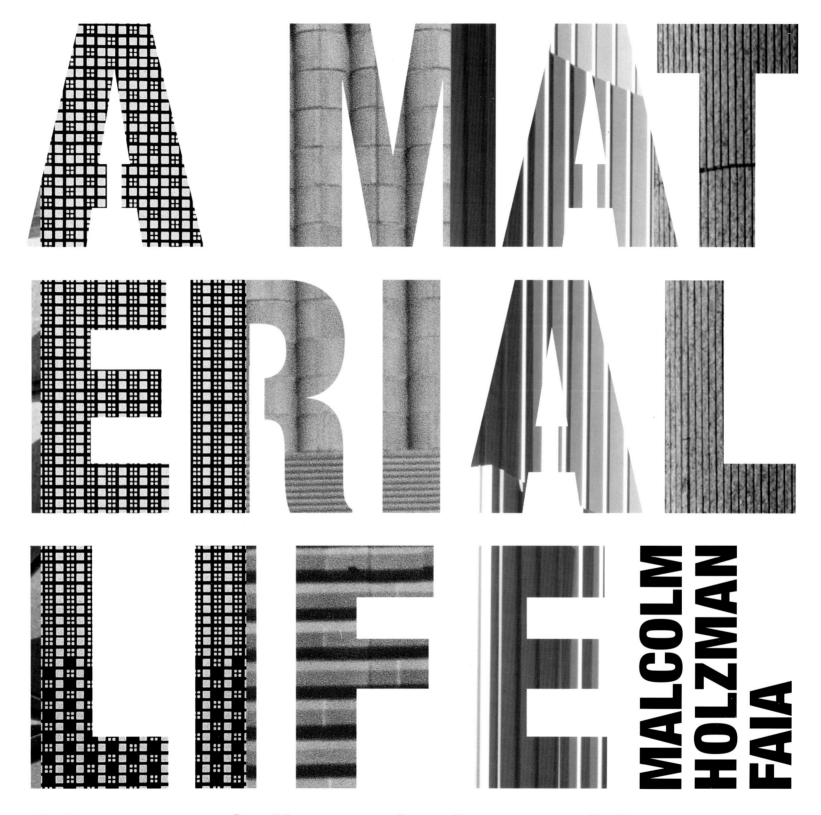

A MATERIAL LIFE

MALCOLM HOLZMAN FAIA

Adventures & discoveries in materials research

edited by Debra Waters

Published in Australia in 2008 by
The Images Publishing Group Pty Ltd
ABN 89 059 734 431
6 Bastow Place, Mulgrave, Victoria 3170, Australia
Tel: +61 3 9561 5544 Fax: +61 3 9561 4860
books@imagespublishing.com
www.imagespublishing.com

National Library of Australia Cataloguing-in-Publication entry:

Author: Holzman, Malcolm, 1940–

Title: A material life / author, Malcolm Holzman.

Publisher: Mulgrave, Vic.: Images Publishing Group, 2008.

ISBN: 9781864702118 (hbk.)

Subjects: Building materials. Architecture.

Dewey number: 720.28   2155

Coordinating editor: Robyn Beaver

Designed by The Graphic Image Studio Pty Ltd, Mulgrave, Australia
www.tgis.com.au

Digital production by Mission Productions Limited
Printed by Everbest Printing Co. Ltd., in Hong Kong/China

IMAGES has included on its website a page for special notices in relation to this
and our other publications. Please visit www.imagespublishing.com.

Foreword 6

01 Theme and variations 8

02 Starting out 18

03 Glazed tile 28

04 Glass 50

05 Metal 76

06 Against the grain 98

07 Shaping clay 116

08 Appropriating materials 136

09 Sustainability 162

10 Art and architecture 176

11 Awareness is everything 199

Appendix 206

Project notes 208

Collaborators 225

Photography credits 227

Biography 228

Contents

Materials play a significant role in architecture. It is important for architects to assume the responsibility for the research needed to locate and apply materials in unique ways. With heightened concern about the use of natural resources, materials for buildings are once again at the forefront of discussion. I strive to take advantage of the variety of products available today, whether time-tested, tired from overuse, relegated to the scrap pile, entirely new, or from allied fields.

The individual sections in the book, organized under general headings for a type of material, describe just a few of the particular actions related to the use of a single product rather than an entire building project. I chose salient aspects of projects that are representative of the research and construction process undertaken to ensure a given material's proper use. A concise description for each building project occurs in the Appendix.

Five years ago in *Stonework,* I illustrated the unique role that stone plays in making exceptional buildings. The positive response to that book led to *A Material Life,* a companion to this previous endeavor describing many of the other products that enliven my architecture. This collection of narrative incidents records the special place materials occupy in my work. The reader may refer to the earlier publication for a fuller description of stone use.

It takes an infinite number of activities coming together to make a book. The untidy circumstances from which this publication springs remains almost unexplainable. The occurrences surrounding the process are presentable, but the actuality of the activities cannot be wholly described since they occurred in and around the daily operations of an active architectural practice.

This book is possible through the tireless and co-operative efforts of the individuals in the offices of Holzman Moss Architecture (2004–) and Hardy Holzman Pfeiffer Associates (1967–2004). I could not have realized all the documented work without the interest, encouragement, and support of my current partner in the practice of architecture, Douglas Moss and my previous partners Hugh Hardy and Norman Pfeiffer. Many individuals at both offices worked on the development of the building projects illustrated. Their support has kept this method of work and research about materials alive. In particular, Steve Benesh, Eddie Kung, Brad Lukanic, Michael Connolly, Nestor Bottino, and Robert Almodovar have played a significant role in developing material applications. A list of the individuals associated with the illustrated projects is provided in the Appendix.

I am appreciative of a number of organizations and institutions that allowed me to present parts of this text during public presentations. These early readings enabled me to clarify the story I was about to commit to writing:

American Institute of Architects, Annual Convention, May 1998. 'Marble Aesthetic on a Dryvit Budget: Materials Use and the Customization Process for the Cost Conscious.'

City College of New York, School of Architecture, April 2004.

SFT Forum, Munster, Germany, March 2005.

The Architects and Designers Building, April 2005. 'The Collaboration of Architecture with Art: A Creative Dialogue,' Malcolm Holzman, Jack Beal, Tom Otterness, and Albert Paley.

American Institute of Architects, El Paso Chapter, October 2005.

I am indebted to Debbi Waters for her stalwart support over the last two years. She reviewed every aspect of this undertaking from its inception to final publication. Her involvement with editing, photography selection, design layout, and negotiations with the publisher were irreplaceable. Especially appreciated was her reading of the text in a number of draft forms and the insightful comments that followed.

Henry Holtzman has again contributed his visual skills to the layout and concept for the design and formatting of the book.

I was fortunate that Jessica Blum, who coordinated photography for *Stonework,* returned to undertake the same role for this publication in addition to chasing down obscure facts, details, and elusive images.

I thank Susan Packard; she saved me through editing and clear thinking from errors in style and language in the first draft of the text.

Many of the suppliers of materials for my projects verified points about specific installations. The assistance of Tom Sawyer, Maury Cullen, John Oberman, Larry Clemens, Lee Olsen, Katie Hinkle, Pete Pederson, Anne Shade, and Verne Larson were instrumental in confirming information that in some cases was a decade old. The artists Jack Beal, Sondra Freckelton, Tom Otterness, and Albert Paley were helpful in recalling the collaborative work process.

This publication would not have been possible without the interest of Paul Latham and Alessina Brooks. I saw little purpose in repeating topics covered and illustrated in the last IMAGES monograph, *Stonework.* Paul encouraged me during many visits to my office to prepare the most exciting book possible.

Malcolm Holzman, FAIA

Star Dust

By HOAGY CARMICHAEL

Slowly

01 | Theme and variations

The materials in my building projects are like the notes on a staff in a musical composition. Themes appear and reappear in variations, like jazz riffs. *Stardust* or *Smoke Gets in Your Eyes,* popular music standards, motivate jazz musicians to amplify these well-known melodies to reveal their distinctive qualities while adding their interpretations and insights. They have proven suitable to myriad artistic representations by innumerable musicians. Materials play an analogous role in my buildings. I examine accepted notions about products to find inspiration for new applications; improvisation can be as much a part of architecture as of music. Materials should serve as a point of departure to identify the essential characteristics of a building and to stretch the boundaries of architecture.

Architecture for me is not about concealment but rather about divulging its very nature to the widest possible audience. Materials are not a mystery; they are an essential building ingredient, our heritage, and part of our everyday lives. For the uninitiated individual architecture can be impenetrable, involving an unfamiliar history, unknown practices, and arcane technical expertise. I use products as observable parts of my buildings to provide identifying keys to the architecture through the choice of the material, surface texture, patterns, and color. I often convey the permanence of the institutions located within civic structures by using stone, a material that can last for centuries. In buildings intended for young people, selecting vibrant colors can enhance lively activities. There is no intellectual or psychological barrier to observing materials. They are accessible, in many cases providing the initial understanding to the architect's intentions.

I avoid limiting my building projects to just a few materials; rather I employ a range of items, exploiting traditional natural materials while making the most of new machine-made products. My interest is in a material's tangible qualities and intrinsic values—grain, color, transparency—not

its reputation as high-tech, primitive, or ordinary. Materials can enhance the forms and spaces they define. One product placed next to another either simplifies or intensifies a design.

Materials can be modest or grand and can even be both at the same time. Every product can call forth an emotional response. Only the designer can conjure up these characteristics to give a space or form specific substance. The deliberate application of materials and the use of light heighten the appearance of my buildings and display their very nature.

Above: Curtain wall, precast concrete column cover, and metal soffit, Performing Arts Center, Texas A&M University-Corpus Christi, TX.
Following pages: Alaska Center for the Performing Arts, Anchorage, AK.

When light illuminates an edge or a glistening surface, it comes to life through contrast, reflection, shade, or shadow while evoking a sense of place. Materials are inseparable from a building's geometry, dimensions, intersections, and a host of details. After carefully choosing the first product, the roles of other materials are determined, necessitating subsequent choices that are compatible or in contrast with the first selection.

Products gain their reputation from their use. Once a material has been used in a particular way often enough, many architects use them in that way only, purely out of habit and tradition. They carry along from project to project like flotsam in a stream. Repetition provides exactly the same results in building after building while interest and spontaneity are lost along the way. Ubiquitous construction materials, satisfactory in almost any situation, are the preference of many architects. Designers tend to follow the conventions of their era, choosing from a cluster of popular items with little consideration or research expended. For instance, glass in all its permutations as an exterior building enclosure and gypsum wallboard as interior space dividers are everywhere today. Because manufacturers provide the data for

construction, product research is not demanded of the architect and even less time is devoted to developing special detailing. Whenever materials selection or other construction factors cause a project to exceed its budget, value engineering is instituted, a process that can degrade materials to reduce cost. Researching materials is rarely straightforward or simple, but it often results in achieving increased construction economy with appropriate product applications. The importance to the owner of a well-researched product accrues in less expensive, more distinctive projects that are of greater value. The cost to the architect for the research is time.

Instead of conforming to the common use of popular products, I approach material selection with complete freedom. New possibilities can transform past restraints. The challenge is to envision the potential of a material, to see opportunity beyond present choices. It is important to look forward and move past clichés that marginalize curiosity and increase dependence on indistinguishable applications. Burnishing an exhausted product can allow it to acquire new distinction. Compelling materials to interact can form new dialogues.

Teasing out rewarding economic results from underutilized products can be as deceptively simple as modifying common items for new uses. Materials can be inspirational; their use often is not.

Practical results may not immediately accrue in proportion to the time spent in materials research and analysis. A product may turn out to be just another version of an existing one. Occasionally, at the end of a careful examination there is only a flicker of an item's viability as a construction product. Unlucky circumstances or lack of a method for application may necessitate putting a product aside for the time being, stowing it away in the unconscious, where it can percolate and spring to life later, when needed.

A standard material can impress me if a hidden aspect emerges. An out-of-vogue product can change appearance with simple adjustments, such as a shift in scale, patterning, or method of installation. I search out new products and materials from allied fields for the broad implications they offer. I look at older structures to provide new insights about traditional products and applications. Taking samples from manufacturers, a time-honored practice, allows me to ponder new uses in the confines of my office.

The ability to employ products economically and with a view toward sustainability and enhanced performance is contingent upon understanding the raw materials and their production processes. Locating the product, investigating its physical properties, developing appropriate applications, erecting mock ups, and overseeing installation come with this design terrain. The reward for all participants is a building that appears rich and vivid while achieving value—both financially and environmentally.

Opposite: Marble quarry, Carrara, Italy.; *above:* Holzman Moss Architecture pilgrimage, Cadillac Ranch, Amarillo, TX.

I anticipated challenges when forming a new architectural practice, Holzman Moss Architecture, with Douglas Moss and ten staff members when Hardy Holzman Pfeiffer Associates, my first firm, dissolved in August of 2004. There were normal concerns: office set up, selecting telephone and computer services, choosing an employee health plan, and so on. However, the unexpected always proves most exhilarating and demanding. This group of founding staff members developed the designs and oversaw the construction of many of the projects illustrated in this book. Working in small teams during the previous decade, they honed the approaches to developing design alternatives with full understanding of budgetary constraints and the selection of appropriate materials. By the end of 2005, the office staff had doubled in size to meet the design needs of our clients. New architects with varying levels of experience joined the original group of individuals to participate in undertaking the work. The unanticipated challenge within the first 16 months of operation was instructing new staff in our well-established methods of design, especially the use of materials. Talk and drawing have their limits; there is no substitute for first-hand knowledge.

Douglas and I decided that something extraordinary needed to occur to speed the information exchange between new and old staff and that everyone needed to partake in this transference since they were ultimately the individuals producing the work. Slowly but purposely an itinerary for an architectural pilgrimage unfolded. Just as breaking the everyday office work routine proved beneficial at the outset of my career, we decided that a break to visit projects was needed for a similar but more focused reason. By traveling to north and west Texas, we could tour seven recently completed buildings in just four days. Along the way, we could also visit other significant architectural and cultural projects, inspect a stone quarry, listen to performances, and see exhibitions, all of which would provide perspective and a means by which to measure our own work. A few office members focused on travel logistics: accommodations, places to eat, and invitations to special guests that would allow this educational adventure to become an enjoyable journey for all of us. Others worked on developing a series of pre-tour seminars to illustrate the design intent embodied in each of the seven projects. The entire office participated in a design competition for an official pilgrimage t-shirt.

On December 1, 2005, 25 staff members joined 50 of our clients and friends in Amarillo, Texas, to start the pilgrimage. An afternoon tour of the Globe-News Center for the Performing Arts with a two-hour critique of the work led by those office members involved with the project launched us in fine style. The hardhat concert (a performance for all the individuals working on the project and community donors) followed that evening; listening to the first music presentations in the 1,300-seat hall was the culmination of our first day on the road. Informal discussions along the way with the project acoustician, theater consultant, contractor, and the client provided added insight. Many of the topics discussed the first day were rehashed during the bus ride the following day between stops in Lubbock, Spring Garden, and San Angelo. The differences between various limestone and granite installations, the firm's three different uses of cattle panels, and the application of oriented strand board (OSB) and timber construction were hot topics. Similar interchanges about design took place at intervals between subsequent stops in Fort Worth, Denton, Plano, Frisco, and Dallas.

With every first-time undertaking, there are risks, whether it is visiting steel mills, brick factories, or your own work. In this case, it was a pilgrimage to orient a dozen new office architects to the knowledge acquired by the others through the research, testing, design, and building of special projects with unique material installations. To pit seeing and discussing the finished projects against the cost of transportation and lodging turned out to be not much of a gamble. The expenses were small in comparison to the value of interaction among newcomers, knowledgeable office staff, clients, and friends, and the collective understanding of the accomplishments in these projects. We returned to our New York City office better prepared to work alongside each other with this shared experience.

Opportunities never dwindle, only the designer's ambition does. After many decades of work, I remain determined, amused, and capable of mastering caution to realize projects of distinction. I recognize that art rises above singular forms, familiar customs, and peculiarities of every kind. I also know that it takes understanding to develop a design idea but it also demands vivid insights into materials to allow a design to pack the emotional power to make it memorable. I often recall that Herman Melville once wrote, "It is not on any map; true places never are."

Opposite (left to right): Movable orchestra shell, Globe-News Center for the Performing Arts, Amarillo, TX.; *right:* Globe-News Center for the Performing Arts.

02 | Starting out

By 1967, our third year of working together, the fledgling firm, Hardy Holzman Pfeiffer Associates, had completed several small projects, primarily house and theater renovations. Hugh Hardy, Norman Pfeiffer, and I were schooled in the prevailing academic doctrine of the post-World War II era; we were trained to design buildings that were modern and

rational. Through our work, we discovered that we were neither totally modernist nor rationalists. Constructing buildings freed us from the design restraints of previous generations and widened our perspective and enthusiasm.

Because we did not have sufficient commissions to keep us busy eight hours a day, five days a week, we often took time to visit places of cultural interest. Breaking the daily routine in the exacting discipline of architecture is a useful practice. A century-old horse farm in New Jersey with an indoor, quarter-mile track; John F. Kennedy Space Center's vehicle assembly and launch facilities; and Columbus, Indiana, a museum-like series of contemporary buildings scattered across a small Midwestern community, were some of our widely diverse outings. Returning to the office from these jaunts, we would hold sporadic slide shows for each other and our small staff. A dozen of these trips might occur in a year, and reviews of our joint discoveries became routine at these gatherings. Although I now have enough work to fill my days, sharing noteworthy observations through office slide presentations still forms a key part of the work process at my new firm, Holzman Moss Architecture.

In those early years, sales representatives, vendors, and prospective consultants dropped by our office without notification or appointments, and when time permitted we talked with them. On his third visit the sales representative from the

Bethlehem Steel Corporation, America's second-largest steel manufacturer, understood our interest in the product and our lack of work. He offered to organize a tour of Bethlehem's 124-acre fabrication facility in Lehigh Valley, Pennsylvania, which in its heyday employed 30,000 people. At one time in the late 19th century, it had the largest single industrial building in the world. This was an opportunity to see the first basic oxygen furnace in the western hemisphere, one that used a new method of making steel. The next week we closed the office for an afternoon and drove to eastern

Above: Indoor track, New Jersey.

Pennsylvania. After a ride through the countryside, we arrived at the plant, where we saw crucibles of molten metal, white-hot sheets of steel fresh from the furnace, and numerous structural columns and beams. It was during this tour that I gained an understanding of the value of visiting industrial operations to observe product manufacturing. At the time, our firm's projects did not require structural elements of the size we saw manufactured at the plant, so I didn't think that what we had seen that day would have relevance to our work. However, in a totally unexpected way, it had a profound influence on the appearance of many of our buildings. Exposed structural systems became a hallmark of our work. For many decades, the quiet presence of the structural framework supporting our buildings gave them order, regularity, and decoration.

The firm received its first new building commission in 1967 for the design of a performing arts center in Ohio at the University of Toledo. We had little understanding about how providential this project would prove, in ways not originally anticipated. The dean of the Performing Arts program at the university wanted a full array of teaching spaces plus two performance venues—a concert hall and a theater. As the work progressed, we built ever-larger cardboard models of these auditorium spaces to clarify and refine the design. After a few months' work, we developed a 500-seat thrust theater with eight cantilevered balconies emerging from the walls encircling the auditorium and stage. The concert hall, similar to a small European opera house, sat 250 patrons in boxes surrounding the entire space and rising from a 250-seat orchestra and stage level to the ceiling.

Eventually, the models grew so big that we could stand in them and observe the auditorium seating areas from the performer's point of view on the stage. When our acoustical consultant joined us in a design review of the concert hall

Above: Bethlehem Steel Corporation, Bethlehem, PA.; *opposite:* Exposed structure and catwalk, Courtyard Theater, Plano, TX.

model, we discussed the need for reflectors above the stage platform that would allow performers to hear themselves and distribute sound to the audience. A variety of reflectors in various shapes and sizes were added to the model for consideration. None proved satisfactory, but we did conclude that the reflectors were necessary and should be transparent, potentially glass, so as not to obscure the auditorium ceiling. Using a ready-made product like automobile windshields appealed to us and, not coincidentally, Toledo, Ohio, was the home of Libbey-Owens-Ford (LOF), a major supplier of glass products to the automotive and construction industries. In 1928 it manufactured the first laminated auto safety glass and in 1931 it provided the 6,500 pieces of plate glass that allowed the Empire State Building windows to be installed as part of the one-year construction period.

During the next scheduled Toledo design review, we arranged a visit to the LOF plant to observe the making of glass, from raw material to the finished product. As we walked from the visitors' parking lot to the fabrication building, I saw a spoil pile, in which chunks of irregular green glass gleamed in the bright morning light. Drawn to this pile of refuse, we couldn't resist handling some of the 3-pound blocks. The Coke-bottle green was mesmerizing; we learned that the color came from the iron content in the sand used in manufacturing. As the tour continued, we watched raw materials processed into liquid and rolled out into long continuous ribbons. During succeeding steps in the industrial process, various products were fabricated from the glass sheets, including windshields. As we retraced our steps to the parking lot, we gathered a small collection of the green glass from the spoil pile. For years, these pieces of cullet graced a prominent shelf in our office conference room, generating more inquiries from visitors than any other item. These samples were fascinating and provocative to all and for me a reminder of their potential use as a building material.

Above (left to right): King Paul & Queen Sophia of Greece visiting Libbey-Owens-Ford; Libbey-Owens-Ford Technical Building, Toledo, OH; Reception Area, Libbey-Owens-Ford showroom at Merchandise Mart, Chicago, IL, 1938, designed by Bruce Goff; *opposite:* Thrust theater model, University of Toledo, Toledo, OH.

Having resolved many of the initial design details for the arts center, including the use of windshields as an acoustical canopy, we were prepared to make a design presentation to the building committee. The models were packaged and taken to Newark Airport for an early-morning flight to Toledo. As we watched the cargo-loading operation from the triangular windows of the French-made Caravelle airplane, we were surprised to see our model boxes taken away from the airplane. I realized that the auditorium-model boxes were too big for the cargo opening. Quickly taking a matte knife from my carry-on bag, I joined the baggage handlers on the ground and proceeded to cut our boxes in half, models and all. (In those days, security protocol was less restrictive.) The reconfigured packages finally made it into the cargo compartment, and our trip to Toledo continued. We received strong encouragement and direction from the client to further develop the design, which meant we had to take the models back to New York with us. Their re-packing turned out to be the most difficult part of our presentation.

At last, three months later, it was time to show our completed vision for this project, windshields and all. This time, with models properly sized for shipment, presentations were made to faculty, students, and administration. The design met with approval from these audiences but we were not done. A special request for an additional presentation to a potential major donor for the project was next. Several hours later, he listened intently to all that we had to say and observed all the images we showed but his response was unemotional. The potential donor was polite and apparently not satisfied with the design. A few days later, a school official explained the donor's tepid reaction; he had expected a building design that looked "Greek." We also learned that this individual was a significant supporter of the excavations at the Agora and the reconstruction of the Stoa of Attalos, in Athens, and had specific ideas about the appearance of the new university arts center. Work on our first new building commission came to an abrupt halt.

Although the Toledo project did not result in a completed new building, it was providential because it propelled us in two directions that we vigorously pursued over the next three decades: creating intimate auditoriums and investigating potential construction materials.

Since that time, the explorations of theater environments that promote intimacy between audience and performer have been refined in more than 150 completed auditorium spaces. On most nights in any given year, 100,000 people across America attend events in rooms we designed. Toledo's two small, unrealized auditoriums set us on this course.

Equally important was the understanding that potential construction materials, in this instance steel and glass, were waiting for us to discover them in fabrication plants and spoil piles across the country.

This was further reinforced a decade after the Toledo experience, when an invitation to lecture at the University of Oklahoma presented itself. I bartered this request for a tour of buildings designed by Bruce Goff, the mid-20th-century architect known for his geometric originality and use of materials. I leapt at the opportunity because I knew that seeing the Goff buildings would be a revelation. Among his many talents was a mastery of unusual building products. Visiting the Oklahoma projects far surpassed reading about them in illustrated journals and books; there is no substitute for the real thing. On this trip I saw a great number of his completed buildings, but it was the client tours of the Ledbetter and Bavinger houses that became the highlight of this visit. Although I knew from publications

that Goff had used glass cullet in the walls of the Bavinger house, it was only during the tour that I recognized the specific material. It was glass similar to that encountered in Toledo's LOF plant spoil pile. However, it was several more years before I had a notion of how he might have run across it. From a catalogue of the first Goff exhibition at the Chicago Art Institute (which houses the Goff archive), I learned that he had worked as a designer for Libbey-Owens-Ford from 1936 to 1937. He spent part of that time in Toledo, where he probably found the glass cullet, just as I did three decades later.

Since these early events in my architectural career, I have accumulated images and information about materials and their methods of manufacture. Only recently did I recall that this routine started at the time of the first new building commission at the University of Toledo. Visiting places of interest and recording my observations started in earnest at that time. Over the years this habit has served me well; it has proven more auspicious than the firm's first new building commission.

Opposite top: Concert hall model, University of Toledo.; *opposite bottom:* Glass cullet coping, Price House, Bartlesville, OK, designed by Bruce Goff.
Above: Eugene Bavinger House, Norman, OK, designed by Bruce Goff.

03 | Glazed tile

Early in 1977, a revelatory catalogue arrived in my office. For years I wanted to use salt-glazed clay blocks, slightly smaller than modern concrete block, but I had been unable to locate a manufacturer. Occasionally, during my travels in the Midwest, I would come upon an old commercial or agricultural structure constructed of these blocks. I saw the structures as a record in the landscape, a mismatch between the building's function and a sturdy construction product. The material and structure most often outlived the activity it enclosed. The warm yellow/orange clay color, sealed with a transparent glaze, held great appeal. Inquiring about their availability, usually in the region where I saw them, became a habit. The reason for their demise in the 1960s eventually became obvious—the manufacturing process for salt glazes could not meet government air-emission standards. On a referral, in one last attempt to locate the material, I telephoned Gladding McBean, in Lincoln, California. I learned that it did not make salt-glazed blocks but was the sole remaining large-scale manufacturer of terra cotta in America. I promptly requested a catalogue. It arrived, launching my extensive exploration into the company's clay products. Although a manufacturer of salt-glazed blocks couldn't be found, I inadvertently discovered another under-utilized clay product surviving well past the period of its most intensive use.

In the early 1900s, a national society promoted terra cotta as a lightweight, fireproof, economic alternative to stone, and 50 large-scale manufacturers across America fabricated it. They produced material for several aspects of construction, from the fireproofing of structural steel to the cladding of building exteriors and the decoration of interior spaces.

Gladding McBean's catalogue was unusual because of its utilitarian character. This was not a bound, glossy promotional volume but a series of loose-leaf pages, essentially construction standards and illustrations of the shapes and sizes of the available material. The catalogue offered products dating back to the company's inception in the 1870s. In addition they were acquiring molds from companies going out of business and were offering architects access to these products as well. At the time of my request for information, its architectural business focused on the restoration of older structures. The catalogue illustrations showed hand-made decorative elements and for the first time, mechanically extruded products. The illustrations suggested neo-Classical buildings. The traditional decorative elements intended for use as cornices, balustrades, parapets, dormers, doorframes, and flat wall surfaces didn't gain my attention. I was curious and intrigued by the extruded elements available in a wide range of colors, with a high gloss or matte finish. If these were not satisfactory for a particular application, creating new shapes and colors at a small additional cost was possible. The catalogue was a treasure trove; I felt like a child getting a hand-me-down Tinkertoy set. The challenge was finding a way to use these apparently out-of-date clay units in an era when few other architects considered this a viable building material for a modern structure.

When the catalogue arrived the design for the Best Products Corporate Headquarters (see pages 56–57), in Henrico County, Virginia, was underway. One early concept for the building's two-story front façade was a large, gentle arc echoing the shape of the interstate cloverleaf it overlooked. Glass along this surface provided natural light to open office spaces beyond. Terra cotta, with its small incremental pieces (readily available units were not more than 24 inches long), would easily lend itself to forming segments of the arc and would be ideal for making the solid portions of the façade, a base, and cornice. The modest size of the terra cotta elements suggested that the glazing could also be in small units, laid up in a masonry fashion. Glass block was an obvious choice. Because color was a benefit of using terra cotta, I selected two standard variations of a light turquoise, compatible with the glass block's greenish tint. The California plant shipped color samples to our office for review. From this exchange, color selection, glaze, and shapes were chosen for a cornice, a base, and coping for a short wall that formed a watercourse 8 feet from the curving building façade. A mock-up of the wall consisting of terra cotta and glass block, erected at the building site prior to construction, was a means of

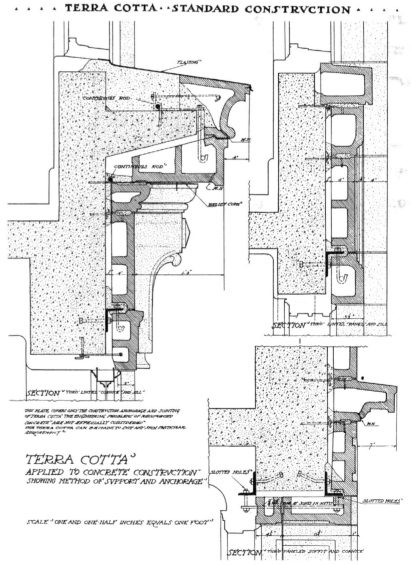

Opposite: Catalogue cover, Gladding McBean & Co.; *right:* Typical detail for terra cotta installation, Gladding McBean & Co.
Following pages: Galbreath Pavilion at the Ohio Theatre, Columbus, OH.

establishing an acceptable quality for workmanship. It ensured the proper results for the finished building, since contractors in this part of Virginia had not installed terra cotta for many decades. Mock-ups, a standard requirement for my projects, guarantee the proper final construction results.

Working with Tom Sawyer, manager of architectural products at Gladding McBean, was both enjoyable and productive. He was conscientious in overseeing manufacture of the product and ensuring a satisfactory installation. He visited both my office, in New York, and the Best construction site to confirm that the workmanship matched the quality of the material. This experience provided a level of confidence for each of us—that Tom could produce the desired building components and that I would use the material appropriately. Given the felicity of both material and manufacturer, I went on to consider both for other projects.

In the early 1980s, a project for the expansion of the Ohio Theatre, an old movie palace in Columbus, Ohio, was in my office. It had actually been there for almost a decade awaiting funding. This 2,700-seat auditorium was a landmark structure by Thomas Lamb, the New York architect known nationwide for movie theater design. The Ohio Theatre was built in 1928, the heyday of movie-palace architecture. Its non-assertive, neo-Classical façade was fitting for its position, across the street from the Ohio

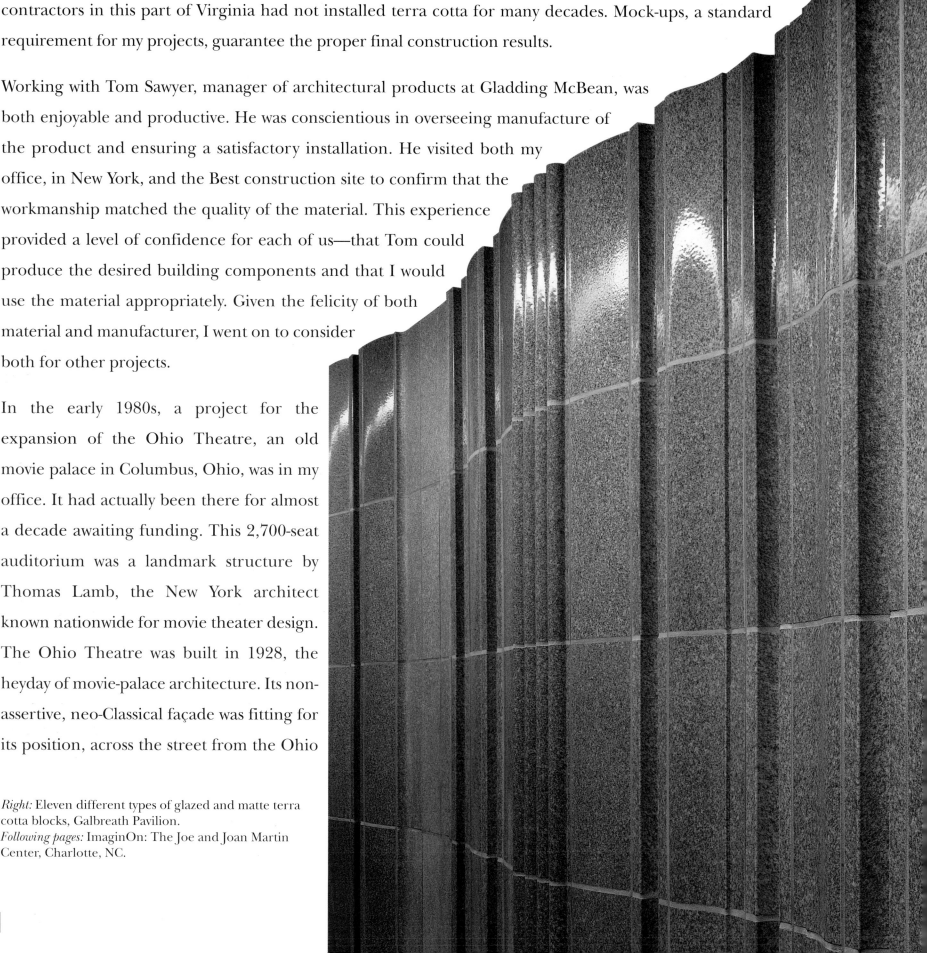

Right: Eleven different types of glazed and matte terra cotta blocks, Galbreath Pavilion.
Following pages: ImaginOn: The Joe and Joan Martin Center, Charlotte, NC.

State Capitol, but the building interior was anything but staid; at the time of construction, Lamb described it as "lavish Spanish-Baroque." It is, in fact, one of the most ornate theaters of the era, with mirrors, gold leaf, chandeliers, decorative plaster—the full works. At the outset of this reuse and restoration project, some members of the Columbus community, among them prominent civic leaders, didn't think much of this tired but flamboyant building. It took many years to secure funding for its conversion into a performing arts center.

Due to its movie-palace origins, the theater's public spaces were small, designed for quick entry and egress between shows, not for gathering at intermissions and before presentations, and the backstage area was practically non-existent. Our assignment was to double the size of the stage, provide an orchestra pit and rehearsal room, expand the public space, and add a lobby. The lobby addition would double as a revenue-producing space for meetings, lunches, and modest community events. The only available land to accommodate the expansion ran the full length of one side of the theater and along a similar narrow area behind the stage.

There was little need to replicate Lamb's design, but there was a desire for a similar sense of luxury and opulence in the new work. The focus of the lobby addition was an existing multi-story firewall that previously separated the Ohio Theatre from the adjoining Grand Theatre. With the demolition of the Grand, this firewall took on new visual prominence. Envisioned as the backdrop for a new ceremonial stair, it rises to all levels of the building. Visible from the exterior through a new full-height glass wall, it is the largest single surface in the addition. The terra cotta chosen to reclad the wall allows it to remain prominent, but also to gently undulate, and with proper lighting, to reflect light in a manner reminiscent of Lamb's opulent original gold-leafed, neo-Classical confection.

Once again, I called Tom Sawyer and reviewed the Gladding McBean catalogue to determine which pieces would best suit this installation. To accent the height of the wall, I decided to turn the terra cotta blocks vertically, allowing highlights to form lines running from top to bottom. This meant the sections originally intended for cornices and moldings would be stacked end-to-end vertically instead of horizontally. A series of red hues were selected to relate to the opulence of the theater's interior and to make the lobby addition a warm environment. Through conversations with Tom, I began to examine the possibilities of a three-nozzle applicator, originally developed to mimic the appearance of granite, which gives terra cotta surfaces a mottled appearance. The applicator sprays the clay surface with a ceramic mixture of frit and metallic oxides, which, during the firing process, develops the desired color and glaze. Rosy-pink and white glazes provide an uneven tone across 4,263 pieces of machine-extruded material on 11 different shapes for this sinuous surface. After selecting colors, I decided to alternate a series of matte blocks with a series of glazed ones along the length of the wall,

further accentuating the verticality of the surface. Much later in the project during the selection of final furnishings and finishes, I returned to the Gladding McBean catalogue for a particular product. A standard item dating from earlier in the century was still available: an ash urn. I selected several 'sand jars,' in the same glaze colors as the terra cotta wall surface to grace the new lobby space.

The favorable outcome of this second project with Gladding McBean prompted a trip to the manufacturer. Observing the manufacturing process is essential to understanding the material and how best to use it. One bright fall day in 1983, my wife and I found ourselves traveling north from Sacramento toward Lincoln, the factory's home. The valley was awash in shades of ocher and umber, evoking the color of clay used for making terra cotta. Like many manufacturing plants across the country, Gladding was situated near the source of its raw products, in this case, clay deposits.

I was surprised to discover the vastness of the operation, a considerable amount of which was devoted to the manufacture of clay pipe. The storage yard, visible from the roadway, contained flanged tubular sections stacked almost 20 feet high. Architectural roof and floor tiles were also very noticeable. It appeared, from the quantity of these materials, that 70 percent of Gladding's business was in non-architectural products. I later learned that one of Tom's responsibilities was to promote the use of terra cotta for new building projects. Following introductions to plant personnel, we toured the "bone yard," where overruns and seconds were stacked everywhere. Best Products and Ohio Theatre remnants were next to replacement pieces for Louis Sullivan's Carson Pirie Scott and Co.'s department store in Chicago. As we walked among these randomly placed objects, Tom confided that the color selected for the Ohio Theatre had been difficult to produce, accounting for the large number of pieces in the bone yard. At the time of manufacture, Tom did not convey the difficulty they had controlling the color. Cadmium red requires a slow firing process at a temperature of 2,100 degrees Fahrenheit. Almost 50 percent of the pieces manufactured ended up in the bone yard because the bottom of the blocks turned off-white instead of red. Since learning this, selecting appropriate colors for firing at high temperatures is a criterion for working with terra cotta.

Above: Boneyard, Gladding McBean & Co.; *opposite left:* Tunnel kiln operation.*; opposite right:* Sample, belt course tile.

In the fabrication sheds we observed raw clay and minerals being prepared for shaping. Plasticity in the clay and fusible minerals are essential to producing a finished, homogeneous product strong enough to carry structural loads. Vertical extrusion presses produced some units, though most production occurred via a horizontal die resulting in "slugs," sections of clay about 30 inches long. After they are dried, glazed, and prepared for firing, these pieces are placed on rail cars that move on tracks through a tunnel kiln for a seven-day firing process. Firing occurs by gradually raising the clay's temperature to 2,100 degrees Fahrenheit. The material is then slowly cooled to

normal temperature. Controlling the heat and its evenness is paramount to ensuring proper shrinkage of the material and the finished product's quality. Four tunnel kilns were operating and the adjacent sheds were hot. In one, skilled workers prepared hand-made sections, those not easily extruded or pressed. A nearby alcove was devoted to historical images and pieces from special projects.

This was an energizing and provocative introduction to the man-ufacture of terra cotta. As we left, I took a few samples from the bone yard, a habit I have acquired over the years. Having representative pieces about the office serves as a reminder of what is easily obtainable. A 2-by-5-inch, golden-colored piece caught my eye. Originally, it served as a section for a traditional belt course. It had three half-round shapes protruding a half-inch from its otherwise flat, mottled, glazed surface. A standard floor tile also provoked my curiosity, because its back side was more appealing than the finished surface. Fine ridges ran the length of

the back of the tile to facilitate the adherence of mortar during the installation process. Made in pairs and broken apart after firing, the edges of the separated surfaces showed a blackening, similar to raku pottery. This "flash" was attributable to their pairing and vertical positioning in the kiln. The irregular coloration was a natural result of the firing process. These souvenirs joined a decorative, Louis Sullivan, terra cotta replacement block from the overrun for the Carson Pirie Scott & Co. department store, as a memento of our visit.

It didn't take long to discover a use for the flashed floor tile. My firm was moving to larger office space in the Flatiron District in Manhattan. The elevator doorframes and adjacent wall surfaces in the new space showed the distress of 60 years of previous usage and repair. This was the ideal location for the floor tile. Installed with the finished side and even color toward the wall, and the ridged, flashed back side facing out, and stacked in a running vertical bond pattern, it added an earthy tone to the new two-story office entry.

Eventually, an adaptation of the golden belt course sample, which I had taken from the factory, found its way into the façade of the Robert O. Anderson Building at the Los Angeles County Museum of Art (page 58). During design development of the museum's Wilshire Boulevard façade, it became apparent that a terra cotta accent for the wall's shallow, three-dimensional tartan-like pattern would enliven the overall glass and stone composition. The wall moved back and forth from its main plane by 4 inches. This dimensional shift occurred by using angled glass block and stone. The terra cotta gave a distinctive inflection to both of these materials; its shape followed the shift in the walls and gathered highlights and shadows in the hazy California sun. The souvenir belt-course tile inspired a new shape—three half rounds at the top and bottom, enclosing a recessed curve. A speckled green glaze influenced by the tile color on the 1931 Wiltern Theater, a neighboring movie

Opposite: Flashed tile, Hardy Holzman Pfeiffer Associates office, New York, NY.
Right: Extruded block inspired by belt course tile, Robert O. Anderson Building, Los Angeles County Museum of Art, Los Angeles, CA.

palace and office tower on Wilshire Boulevard, was chosen to contrast with the yellows and pinks in the museum's limestone. This terra cotta unit, though it had similarities to pieces illustrated in the Gladding McBean catalogue and the souvenir sample, required a new extrusion. The large quantity of units ordered offset the minimal cost to manufacture a new die for this special shape.

Tom and I worked on subsequent projects with considerable success, one of which was pure pleasure. In 1987, my wife and I moved to a loft, a commercial space on the top floor of a 100-year-old building near the well-developed Manhattan neighborhoods of Gramercy Park, Madison Square Park, and Murray Hill. There was no need to restrain the use of materials, as they only needed to meet our standards of acceptability and budget. Because the loft's distressed appearance appealed to us, we limited improvements, allowing the sense of the raw commercial space and its history to remain. We stained the maple floor dark red to disguise scarring from extended use and abuse, stabilized the edges of the missing pieces of the plaster ceiling, and painted the original steel windows.

When it was time to discuss the bathrooms, we recalled from our visit to Lincoln that Gladding McBean had the ability to prepare almost any color as a finish and that samples abounded. During our plant tour, we had seen dozens of closely related glaze samples for each tile run. The need for slight tonal variations resulted in excess tiles. The selected sample was a guide for the manufacture of the finished product, while the others ended up in a large pile designated for eventual recycling. Some discarded samples showed small areas removed at the center of the tile—test patches to determine glaze thickness. Others were deformed from the glazing and firing process, while still others were stacked and covered with dust and grime from the surrounding operations. When considering our bathrooms, we recalled these 30,000 samples. They were 5 by 2.5 inches and less than half an inch thick. Tom, perplexed by our desire to use these mismatched samples, was, after numerous conversations, persuaded to forward 7,000 remnants. My wife and I set to work, installing them to show off the variety of colors. A vivid and distinctive crazy quilt to live with, the tiles are a daily reminder of completed projects and avoided color choices.

The largest terra cotta undertaking I ever contemplated was part of a proposal for the expansion and renovation of the 1904 Willard Hotel in Washington, D.C. Known as the hotel of presidents, it figured prominently in the city's political life for many decades. Intense use of its public spaces at the beginning of the 20th century gave rise to describing individuals striving to influence government as lobbyists. The building was a blight on Pennsylvania Avenue for a number of years before the Kennedy administration took office. The appearance of this abandoned hotel during Kennedy's

Opposite: Installed terra cotta paint sample tiles, Landsman Holzman Residence, New York, NY.

inaugural parade led to formation of a commission to restore the entire avenue, with the Willard its keystone. A Florida commercial developer, a California hotel operator, a New York architect, and a Washington contractor made an unlikely consortium, winners of the 1979 design competition. This was the first project at this scale in the Postmodern design era to suggest xerography as an approach to design. The intention was to expand the existing hotel by surrounding a public courtyard facing Pennsylvania Avenue with four smaller replicas of the original Willard. The significant difference between the historic structure and the additions was the contrast of new, glazed terra cotta façades with the 80-year-old

stone and brick walls. The project design was the toast of the town for six months, but soon suffered from soaring interest rates and difficult economic times. This national economic recession caught the developer under-funded; another development team completed the project. A semblance of the original design was constructed, but without the shimmering glazed surfaces originally envisioned.

The office and loft experiences with terra cotta served me well in the next generation of work. In 1988, Tom Sawyer became part owner of the Quarry Tile Company, a tile manufacturer in Washington State. This operation was different

Above left: The John F. Kennedy inaugural parade passing the Willard Hotel.*; above right:* Additions to the Willard Hotel, Washington, D.C.
Opposite: Restoration of The Pride of the Pacific, Hawaii Theatre Center, Honolulu, HI.

46

from Gladding McBean; it produced flat materials for wall and floor installations, not three-dimensional blocks. He had barely taken residence there when I imagined that glazed tile would make a good wall surface in the corridors outside the dressing rooms of the Hawaii Theatre, in Honolulu. The last movie palace in the downtown, it was known locally as "the Pride of the Pacific." While working on its design and restoration, I thought tile could enliven the basement circulation spaces. An intense color would accomplish a lot in this heavily trafficked actors' space. I worked with Tom, using the knowledge secured from reversing the tiles in my office, to produce a new, grooved flat tile. Unlike the floor tiles, these had regularly spaced slots stamped into the surface during the manufacturing process. From a few samples, we developed a 6-by-6-inch cobalt blue tile for the dressing-room areas. As part of current public backstage tours of the rejuvenated theater, a regular stop occurs at the tiles. Visitors are told that it is well known in Hawaii that *obake* (ghosts) don't like water; the color blue scares them away.

Incorporating the same tile shape in a number of other colors became an office practice. The entry pavilions at the Berrie Center for Performing and Visual Arts at Ramapo College, in Mahwah, New Jersey; the light and sound locks to the auditorium at the Lied Education Center for the Arts at Creighton University, in Omaha, Nebraska; and the surfaces in the food-serving area of McClurg Hall at the University of the South, in Sewanee, Tennessee, all benefited from the use of this material. Variations on this tile became an office convention within three years of its initial manufacture and application.

Opposite: "Obake" tile, Hawaii Theatre Center.; *above:* Lobby, Lied Education Center for the Arts, Creighton University, Omaha, NE.

As Tom's equipment became more sophisticated at the new operation, he developed innovative product lines with water-jet and laser cutting equipment. Making tiles in different colors and interchanging pieces from one to the other, thereby getting two colors in one tile, became one of Tom's passions. The very first tile done in this technique was so corny that I saved it. Two 6-inch-square tiles, one blue, the other white, were laser cut with a Scottie dog at their center. The dog was chosen to illustrate the refinement that could be achieved in depicting its fur coat. The centers were then exchanged to provide a dog on a contrasting background. I employed the contrasting color technique in one tile application, without Scottie dog, in the colorful curving walls on the main and lower levels of the Louis Stokes Wing of the Cleveland Public Library, in Ohio. Here, the conjunction of laser-cut tiles with shaped and flat tiles animates a vividly patterned 130-foot-long wall, which runs from the main entry to the primary vertical circulation points and contains the circulation desk. It is one of the library's defining elements. The shape of the wall and its bright hues—yellow, orange, blue, and green—provide a lively welcome to all library visitors regardless of the local weather, which is frequently gray.

Finding new uses for standard materials, employing age-old products for contemporary efforts, and developing new ones, provide personal satisfaction and public delight. Exploiting changes that new technology brings to the plasticity of clay is equally invigorating. I take pride in both the ability to use clay properly, which ensures lasting results, and my familiarity with the product, which allows for a variety of new applications.

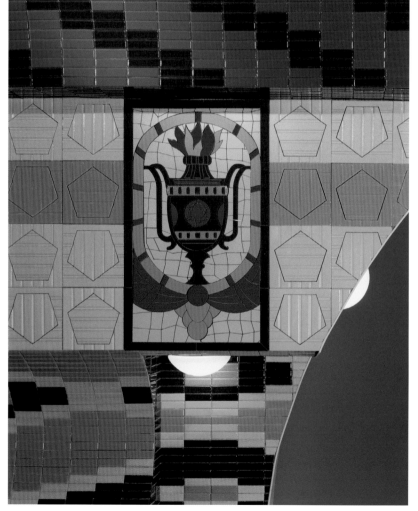

Above: Scotty dog tile sample.; *right:* The *Light of Learning,* Louis Stokes Wing, Cleveland Public Library, Cleveland, OH.
Opposite: Tile wall, Louis Stokes Wing.

04 | Glass

My first-year design studio instructors at Pratt Institute encouraged us to enhance our drawing skills by looking at art, and we made frequent trips to New York museums and galleries. Frederic Edwin Church's (1826–1900) landscape paintings aroused my interest the first time I saw them. My initial experience with Church's paintings was not much different from other observers a century earlier. He dramatized the splendor of the numerous places he visited in the Americas: the Arctic and its icebergs, the tropics and erupting volcanoes, sunsets, and rainbows. His paintings caused a sensation when originally exhibited in New York City, where hundreds of people waited in long lines to see these vivid representations of exotic places. These works were still considered striking in the 1960s, though people didn't wait in line to see them. The natural environment was under siege; physical change threatened the wilderness Church depicted. Initially his images added to the burgeoning public interest in the vastness and beauty of the American landscape, subsequently they represented its shrinkage.

Curious about his work, I visited Olana, his Persian-inspired villa on a 250-acre Hudson Valley estate, 100 miles north of New York City. Church designed this confection of Middle Eastern and Victorian exuberance with Calvert Vaux, architect of the structures in Central Park. Olana, a National Historic Landmark, offers a romantic visual feast: architecture, interior decoration, Church's own painting collection, and hilltop views from windows, porches, and lawn to the Hudson River and the Catskill Mountains beyond. The strong sunlight washing down on the landscape shifted as the afternoon unfolded; the trees, river, and mountains changed from hard-edged objects to soft blazes of light, similar

to a twilight scene from one of his paintings in a distant location. The views from Olana were incandescent, prompting me to reconsider his paintings.

During subsequent museum visits, my focus shifted from the scenes depicted in his paintings to the images' luminosity and their evocation of light. The effects that attracted me were the same ones that made other observers describe them as kitsch. To them, the pink and orange glow of a sunset, the blue-white reflected light from an iceberg, the translucent rainbows, and the vast wilderness and its diminutive people appeared dated. In many instances, museums relegated Church's paintings to obscure gallery locations, indicating the importance placed on them at that time.

The Corcoran Gallery, in Washington D.C., was an exception; it prominently displayed a large, 7.5-foot-wide Church painting, *Niagara* (1857). Even though Niagara Falls was a well-known and frequently depicted tourist attraction, this composition and its light was a revelation to me. Church showed the falls from an unusual vantage point—above. By forcing the observer to look across the rapids and wide expanse of the river surging over the precipice, the roar of the rushing water can easily be imagined. The light in the image illuminates the transparent spray from the foreshortened horseshoe falls.

Opposite: Olana, Hudson, NY.*; above: Niagara*, 1857, Frederic Edwin Church, Corcoran Gallery of Art, Washington, D.C.

I found implied instructions in this painting. The visual entry into this unusually proportioned picture is paramount, setting the stage for everything that follows. Of equal consequence is the luminosity of the image. Light supports the composition, and the variation in its quality carries an emotional charge. The painting devices that Church employed so well made me consider their application to animate my architecture.

Many mid-20th-century architects eschewed the incandescence Church depicted in his 19th-century paintings. The resulting modern buildings were flat, unadorned structures with simplified application of materials; sunlight was not a notable animating factor. I was fortunate; my architectural practice involved the design of public structures, such as museums, theaters, and libraries, which needed to be self-advertisements, displaying activity within. Carefully opening a structure through a partially transparent or translucent external skin and rendering the entire composition in light embodies my aspiration to accomplish in architecture what Church achieved in his paintings, an invitation to the observer to participate—to view something memorable.

In 1976, Hardy Holzman Pfeiffer Associates received its first national American Institute of Architects award, for the small 21,000-square-foot Columbus Occupational Health Center in Indiana. Serving as a clinic for employees of the Cummins Engine Company, it provides for routine medical examinations, testing, and treatment of occupational illness and ailments. Rather than conceal functions such as waiting, circulation, and laboratory activities, the design

Top: View from Olana.; *above:* Columbus Occupational Health Center, Columbus, IN.; *opposite:* Laboratory and waiting area, Columbus Occupational Health Center.

brought the non-medical spaces out into the open. We glazed many of the spaces along the central circulation path, interior rooms that did not require visual privacy, and illuminated them with natural light, thus providing a sense of community and human activity not found at the center of similar structures.

Dark insulating glass encloses the square, two-story building. A diagonal skylight intersects this volume and follows the circulation path from the front door to the internal waiting areas. We minimized construction costs by employing readily available standard curtain-wall and greenhouse assemblies but arranged them unconventionally. In the evening, the majority of the building exterior becomes transparent. Vibrant and warm colors illuminate the interior spaces through the glazing.

In the 1970s, the Pittsburgh Corning Glass Company announced its intention to discontinue production of glass block, a staple of the business since its introduction in 1937. The standard unit consisted of two cast-glass sections, fused along

a common joint, which form a vacuum in the enclosed air space. The product was strongly associated with commercial installations and the remodeling of out-of-date façades. Architects bemoaned the company's decision, noting the importance to their work of this specialized type of glass. This outpouring of sentiment ultimately caused the company to reconsider its decision and reverse its direction, instead studying improvements to its products. The company eventually introduced new glass patterns, surface coatings, light filters, and additional block types. These improvements allowed for better control of light and heat gain, and offered the potential for more varied applications.

The curving façade of the Best Products Corporate Headquarters faces a cloverleaf intersection on Interstate 75, north of Richmond, Virginia. A terra cotta base, cornice, and adjacent watercourse constitute part of its façade. Small glass-block units complement the scale and color of the terra cotta and run from base to cornice while conforming to the arc of the façade. Two newly improved glass blocks with light filters and surface coatings animate the two-story façade that illuminates the open office areas behind. Nearly invisible at one time of day, the large diamond pattern in the wall created by the two types of glass block becomes a reflective surface at other times, while the filtered light in the interior open offices changes during the course of the day.

The Anderson Building at the Los Angeles County Museum of Art creates a different effect with glass blocks. Natural light enters the Wilshire Boulevard façade through a band of glass blocks and bricks, and then travels through a series of internal metal louvers at the top of the gallery walls, providing diffuse light to supplement the special art lighting. Two filters built into the internal air space of the ribbed glass block remove ultra-violet light and assist in distributing it evenly. Stretching standard, hexagonal-shaped, Pittsburgh Corning blocks permitted the plane of the exterior wall to change. Longer than their standard hexagonal corner blocks, normally employed to make right angles at wall intersections, these units allowed the wall surface to angle back 4 inches and to move forward a similar amount. With special terra cotta sections and cut stone forming the same obtuse angle as the glass blocks, all materials follow the same wall contours, a three-dimensional tartan-like pattern. Along the exterior façade, the glass blocks and terra cotta add an accent to the walls' warm-colored stone surfaces.

In the main reading room of the Columbia Public Library, in Columbia, Missouri (page 104), the quality of changing light defines the space. Stacked 10-foot-wide panels of two sizes of translucent glass-blocks frame folded sheets of transparent glass, providing views to the surrounding community. A variety of seating locations allows users of this conically shaped room to focus inward on library materials or, if desired, outward to framed views of iconic community structures like the 1895 Jesse

Hall on the University of Missouri campus, the Boone County Courthouse, smokestacks, and a city water tower. In the evening this space and the lobby below presents the library's two entries as a luminous beacon.

In 1999, during an interview following the completion of a music facility at the University of North Texas, a journalist pushed hard to get a usable quote. In exasperation I blurted out, "Nothing says you have to listen to music in the dark." A few months later, while driving down a Texas highway, I was surprised to hear myself thus quoted on the radio, in an advertisement by the Texas Society of Architects/American Institute of Architects promoting architecture.

In the 17th and 18th centuries, luxurious, light-filled residential salons were often settings for musical performance. In contrast, natural light was excluded from theaters of this era to provide total control of the performance. By the 20th century, natural light was also excluded from music performance halls even as urban orchestras began performing in rural outdoor settings during the summer season. These outdoor performances and reconsideration of the older residential music settings led me to reflect on the inclusion of natural light in new enclosed music halls. Unlike theater spaces, music halls do not require total control of lighting to elicit the correct ambience for the audience; rather, they need to provide a comfortable setting and, above all, good acoustics. In fact, the wandering eye is part of the experience of listening to music. Seldom does an audience member sit staring at the musicians for an entire performance, so providing a sense of the world outside the performance chamber without interfering with the listening experience makes perfect sense to me.

Above: Concert hall, Center for the Arts, Middlebury College, Middlebury, VT.*; opposite top:* Center for the Arts, Middlebury College.
Opposite bottom: Rooftop dormers, Center for the Arts, Middlebury College.

My first opportunity to include natural light in a music performance space arose when I was designing a hall in the Center for the Arts at Middlebury College in 1992. The college needed a small, 350-seat, intimate music room as part of a larger arts center. The campus is just on the edge of a 245-year-old Vermont village, so excluding exterior noise was not a major concern. As the design developed, it became obvious that natural light could enter the chamber without interfering with the performances and acoustics. After exploring various alternatives, I placed 14 rooftop dormers above the performance platform and two above the audience seating. One triangular face of each dormer contains three clear panes of glass, one small blue pane, and an electronically controlled shade. These counterpoints in the large acoustical volume of the space provide glimpses of sky, moving clouds, and stars, a complementary view while listening to music.

The Middlebury endeavor emboldened me to consider a larger installation in another music facility, Winspear Hall, at the University of North Texas, in 1999. The music programs at this institution had been successful for many decades, but the existing auditoriums were inadequate for the performances that took place within them. There was demand for new auditorium space both during the day and in the evening for presentations by bands, ensembles, and choruses in addition to visiting performers. It seemed appropriate to introduce controlled natural light to the main presentation room of the new facility. In music halls, the appearance of the surface directly behind the stage is important because it is the first surface an audience

looks at once attention strays from the performance. For this reason, at Winspear, the pentagonal rear wall became the obvious place for inclusion of natural light. The client enthusiastically adopted the idea; a long period of design exploration followed to ensure that glazing this large surface would meet acoustical requirements. In the resulting design, layers of insulating glass, glass block, and opaline glass filter the light and keeps out extraneous noise. In the hall, sheets of yellow marbleized glass follow the geometry of the end wall. During the day, modulated light from the north causes this glass surface to glow. At night, the glow reverses—artificial light in the space indicates to audience members approaching the building that the hall is in use. The translucent surface provides no view out of the space, but the presence of the exterior environment is inescapable.

The luminance generated by the rear stage wall is similar in intention to the light radiating from the chandelier-like acoustical reflectors in the audience chamber. Seven floating pentagonal disks echo the shape of the rear stage wall while providing locations for performance, house lighting, and reflective surfaces to direct sound out to the audience and back to the performers. A folded polycarbonate panel forms the lower layer of each chandelier, picking up and distributing light from a series of small incandescent bulbs. These objects hover in the space, appearing lighter than the opaque room surfaces enveloping them. They add light for program reading and lend sparkle to animate the large acoustical volume. Consideration of translucent acoustical reflectors began with the design of the University of Toledo Concert Hall and they were first installed in Boettcher Concert Hall, in Denver, Colorado, in 1978. Since then, illuminating these devices to create special effects has evolved to enhance the overall environment while meeting specific acoustical requirements.

The desire of many public institutions to shed the image of the hermetically sealed, strong-box has permeated architectural commissions in recent decades. It has been a slow evolutionary process away from opacity. Improving these

Opposite: The Margot and Bill Winspear Performance Hall at the Lucille Murchison Performing Arts Center, University of North Texas, Denton, TX.
Above: Glass wall detail, Margot and Bill Winspear Performance Hall.
Following pages: Lucille Murchison Performing Arts Center.

spaces' technical environments and showing off the lively activities taking place within them provided the impetus. In the 19th century, top-lit exhibition galleries were commonplace. By the mid-20th century, many of these skylights were no longer effective, but natural light was still unwelcome. By the end of the 20th century, however, restoration projects at museums re-employed skylights, with new technology, to the delight of the public and museum professionals alike. Their enthusiasm accompanied recognition that art objects are best seen in the full color spectrum of natural light. This process carried forward in the design of new exhibition spaces as well as restorations.

In the design of the West Wing of the Virginia Museum of Fine Arts, to house special collections given to the state by Sidney and Frances Lewis and Mr. and Mrs. Paul Mellon, the inclusion of natural light was a major consideration. Mr. Mellon's familiarity with the benefits of natural light stemmed from his involvement with construction of the East Wing of the National Gallery of Art, in Washington, D.C. and other museum projects. He wanted the West Wing galleries to benefit from natural light. Paul Marantz and Richard Renfro, well-known lighting designers, and I worked on the development of this aspect of the project. We studied a number of glazing systems for the modest number of openings in the façades facing the Robert E. Lee Camp Confederate Memorial Park; two held considerable promise. Frit glass, a product used in the automotive industry for sunroofs and windshields was a possibility. At the time, Pittsburgh Plate Glass was developing a process for applying ceramic coatings to glass to increase shading and reduce heat transmission. After careful exploration of the use of perforated metal joined to glass, we concluded that the difference in expansion between the two materials could not be resolved for this

Right: Acoustical reflectors, Boettcher Concert Hall, Denver Center for the Performing Arts, Denver, CO.

installation. We decided to proceed with a conventional approach to the gallery skylights and to employ frit glass on the museum façades.

To avoid the problems of traditional skylights, we developed a skylight system that allowed for rooftop enclosures with mechanically adjustable louvers to control the light levels in the galleries. This system assured that no direct sunlight would enter the exhibition spaces. Many visitors are unaware of this hidden glazing and shading system, since it is concealed above a lower layer of glass at the ceiling, but benefit from the even and changing wash of light in the four uppermost Mellon Galleries. In the largest Lewis Gallery and the Marble Hall of the West Wing, light washes down from a similar glazing and shading system, giving the wall surfaces a warm reflective glow.

The longest façade of the West Wing faces the setting sun, a difficult orientation for light control. Solid limestone blocks pattern 95 percent of this surface. Enlivening and interrupting this 300-foot-long façade facing the adjacent park are two glass enclosed stairs and an exterior sculpture terrace. The stairs connect the two main gallery levels in the building and allow for a break in the flow of circulation between the exhibition spaces. A dense ceramic frit pattern on the glass sheathing the stairs filters out 80 percent of the light, minimizing heat gain and glare while framing views out to the surrounding greenery.

Since this pioneering architectural application for exterior building openings, frit glass and patterned laminated glass have become standard products. Today they reduce glare, provide shading, and reduce heat gain, in installations from commercial high-rise structures to classroom buildings. The ten-story glass oval of the Louis Stokes Wing at the Cleveland Public Library, 1999, and the two-story curtain wall installation at ImaginOn: The Joe and Joan Martin Center, 2005, both employ this type of glass. In these projects, the glazed surfaces exploit the environmental characteristics of the product while using patterns in the glass tailored to the building. At the Stokes Wing, the pattern is dense at the top of the glazing, near the ceiling, and diminishes so that it becomes

Opposite top: Mellon Galleries, West Wing, Virginia Museum of Fine Arts, Richmond, VA.
Opposite bottom: Lewis Galleries, West Wing, Virginia Museum of Fine Arts.
Right: Frit glass stair enclosure, West Wing, Virginia Museum of Fine Arts.

transparent at the level where users of the library look out to the surrounding mall, downtown, and Lake Erie. Interior shades provide additional control at certain locations. At ImaginOn, small ceramic parallelograms form the frit pattern while a series of regularly spaced metal fins attached to the curtain wall provide exterior shading. In each instance, the glass provides visibility into the building to allow the passerby a view of the activities within. Both projects are modest-sized structures in their local environments. Even so, they have become nighttime landmarks because of their luminosity.

Texas A&M University-Corpus Christi started as a local college on Ward Island in 1947 from a former United States Navy Station. Disaster struck the institution in 1970; Hurricane Celia damaged a large part of the facilities and consequently required the administration to institute a new master plan for growth. In 1993, the college joined the Texas A&M system as a four-year comprehensive university.

During the last decade significant building development occurred in response to the growing number of students attending a variety of expanded programs. The number of structures constructed at the edge of the campus adjacent to the Gulf of Mexico has increased. This location on the coastal bend strongly influences construction; buildings face the full force of winds coming off the water. The design of the new Performing Arts Center reconciles the need for safety while taking advantage of the unique views.

The first phase of construction contains a 1,500-seat concert hall and support spaces. A large roof and overhang covers all activities. Textured precast concrete panels, clay tile, and brick walls surround all the building activities except those in the lobby, in this area glass wraps around the eastern façade looking out to the Gulf. Steel trusses incorporated into the curved glass surfaces enhance the strength of the curtain wall system. Special glass coatings and shade from the roof overhang reduce the amount of light and heat transmission to the lobby. During the evening hours, the transparent glass surface makes audience members awaiting the performance on the three main lobby levels visible to others approaching. Unparalleled views to the Gulf are available to patrons from this space. The lobby area will expand along with the construction of a 300-seat theater, education spaces, and additional support spaces in the next phase of construction.

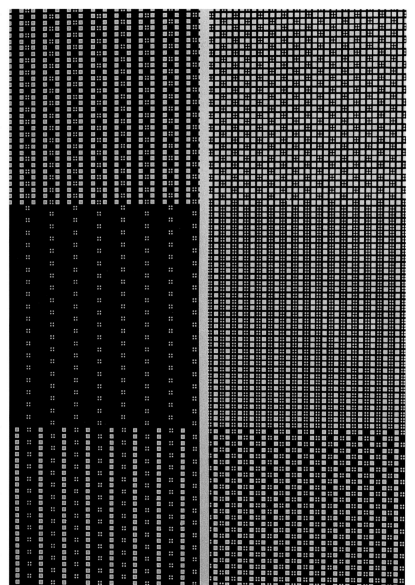

Church's paintings did not mechanically reproduce nature, but rather used the contrasts of light in specific natural settings to evoke emotion and stimulate the viewer's imagination. He employed light to make associations, offering windows to another world. Incorporating glass to dramatize its transparent and translucent qualities juxtaposed with other materials has become an enthusiasm. Glass supplies the counterpoint to the opaque characteristics of many materials and, at night, a transparency reversal due to the source of light offers opportunities provided by no other material.

Opposite left: Louis Stokes Wing, Cleveland Public Library, Cleveland, OH.; *opposite right:* The frit glass, Louis Stokes Wing.
Above: Sample, frit glass.
Following pages: Performing Arts Center, Texas A&M University-Corpus Christi, TX.

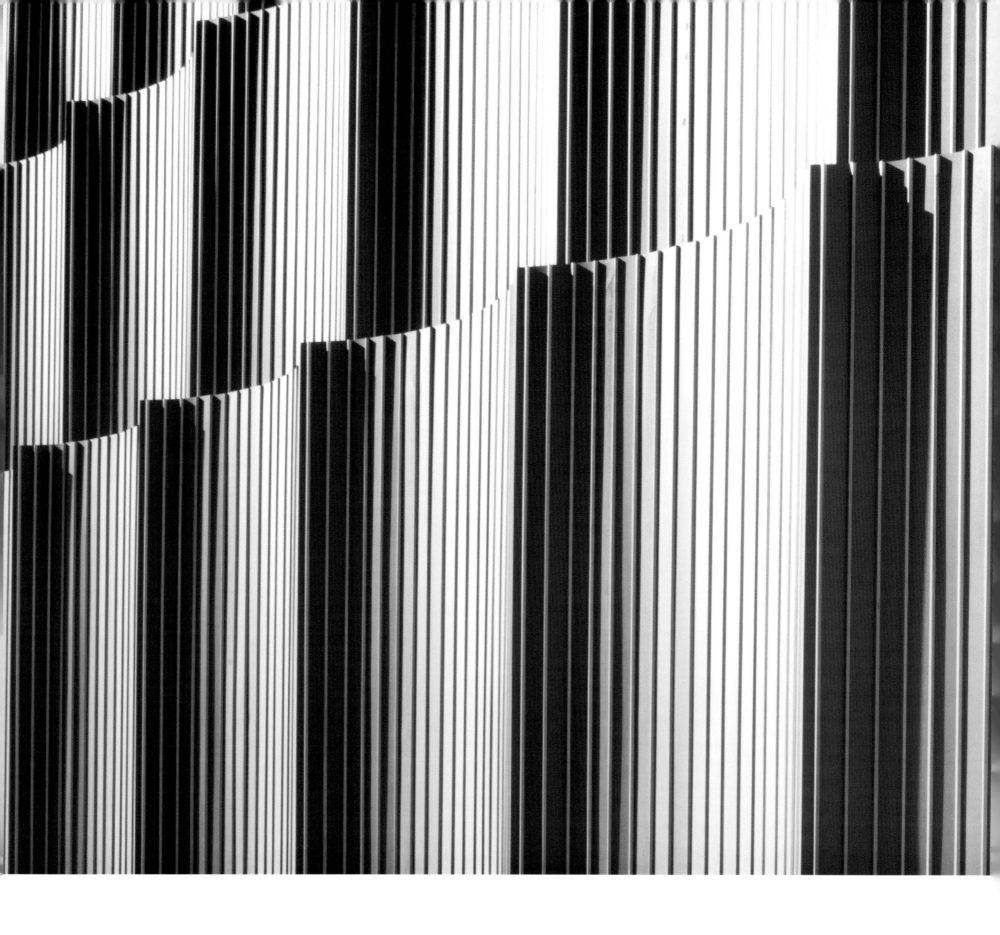

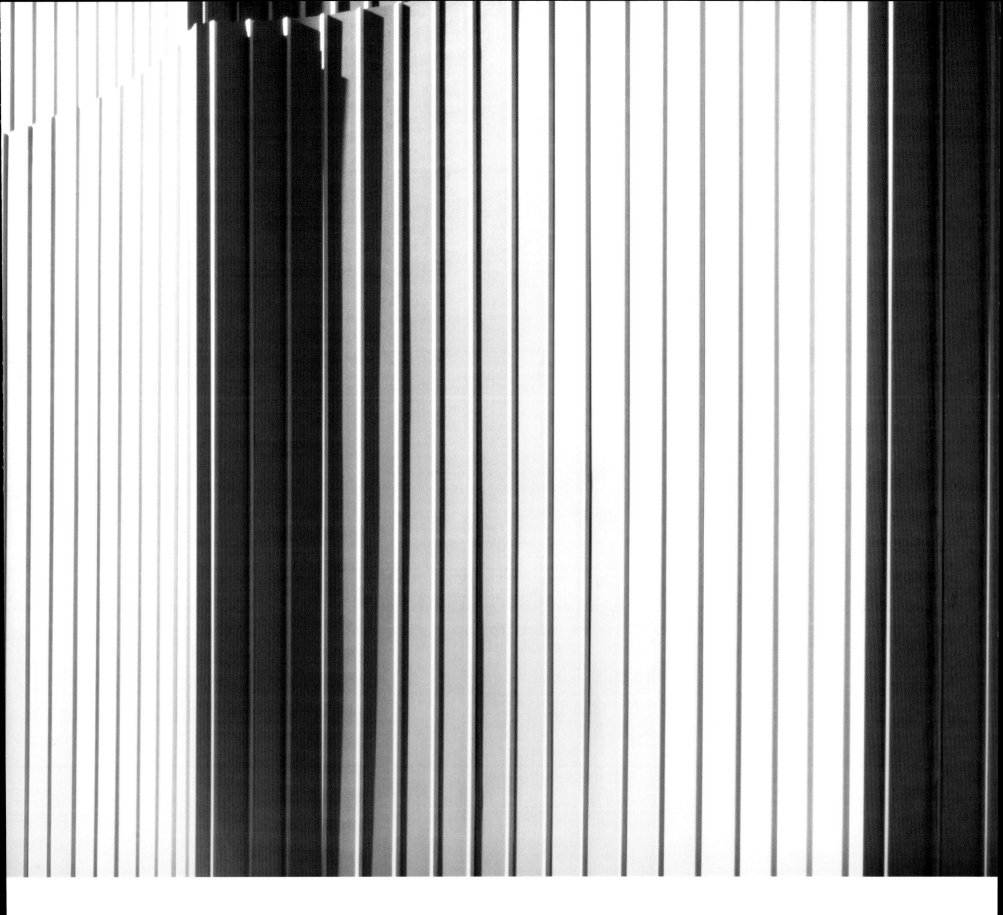

05 | Metal

Revere Copper Products, Inc., founded by Paul Revere in 1801, is one of the oldest continuously operating manufacturing companies in the United States. Mass production of copper products started shortly after Revere's famous horse ride, ending England's control of the American market. Its home, Rome, New York, is a small city in the middle of the state, in a pastoral area of farms and rolling hills. This copper plant, one of the nation's largest, is composed of a group of industrial structures that reside in stark contrast to the surrounding agrarian landscape. At Revere, raw materials, machinery, and skilled laborers bring forth a variety of finished copper products.

On a warm day in the late summer of 2001, I treated myself to a drive through the scenic New York countryside to observe this industrial process. It was fascinating to watch raw materials, heated beyond the liquefaction point (1,981 degrees Fahrenheit), placed in a mold, and then rolled out into a 3-foot-wide, 6-inch-high, 25-foot-long continuous cake, a large ingot. As the cake emerged from the furnace, it glowed bright yellow-orange from the heat. As it reached its full length, the far end had already begun cooling, turning black. Transporting the 99-percent pure, 2,300-pound piece of copper to the next operation on compressive rollers reduced its thickness and elongated its dimensions. In just one hour, other compressive rollers transformed the cake into a shiny copper coil, containing 5,000 feet of 16-ounce (.0216) sheet material. The coil was ready for conversion into myriad new copper products.

The visit to the Revere plant took place after the 38th annual Copper Development Association meeting, in June of 2001. My invitation to address this august group of executives resulted from three decades of using copper products in distinctive ways in a variety of projects. I followed a talk given by a representative of the United States Environmental Protection Agency that outlined directions in recycling and waste management of materials. Images that accompanied my talk showed copper used in a number of architectural installations. One of these, the San Angelo Museum of Fine

Above: Copper coils, Revere Copper Products, Rome, NY.; *opposite top:* Standing seam copper roof, San Angelo Museum of Fine Arts, San Angelo, TX. *Opposite bottom:* San Angelo Museum of Fine Arts.

Arts, has a vaulted copper roof that curves in two directions. Sheathed in a standing-seam application (in which a continuous folded metal joint connects adjacent sheets), diagonal seams run over the barrel vault of the roof, a configuration that took a lot of ingenuity to install. The project received a number of awards; the most satisfying of all the recognition conferred was its appearance on the cover of the local West Texas telephone directory, indicating its place of importance in the community.

Descriptions of my final two projects garnered the audience's undivided attention. In 1978, a fortuitous meeting occurred with Shorty Welsh and Verne Larson, long-time employees of Ettel & Franz, a well-established roofing contractor in Minnesota's Twin Cities. I had been contemplating the use of a standing-seam copper installation on the roof and walls of the WCCO-TV Communication

Center and Headquarters, a regional television station in downtown Minneapolis. During our initial conversation about the project, Shorty inquired about the use of copper shingles. After indicating that I was unfamiliar with them, he offered to fabricate a few and bring them to our next meeting. In the intervening weeks, Shorty cut enough copper from a coil to make a few sample shingles. I discovered that he ran the copper sheeting through a crimper, corrugating the flat surface to make it more rigid. This process prevented the finished sheet goods from "tin-canning," or unintentionally wrinkling. He also folded the copper shingle once to make it fit snugly into surrounding shingles, concealing exposed edges from the weather while covering the fastening devices. This was something I had never seen before, the work of an innovative roofer. The client for this project was also pleased with the sample, so installation of the material proceeded on this landmark Nicollet Mall structure. Today, the green patina of the copper both complements and contrasts with the yellow Minnesota limestone that sheaths the remainder of the building.

After using these shingles on projects from Middlebury, Vermont, to Anchorage, Alaska, I telephoned Vern with a problem. My cabin, a 30-year-old rustic retreat in upstate New York, was wearing out. The asphalt shingles on its small roof had weathered down to their paper backing. I knew that a copper roof would last 80 years. I had no desire to replace this roof again, so I was calling to secure some copper shingles. Ettle & Franz was generous, forwarding 500 square feet of material to me. After a local roofer and I discussed the method for interlocking the shingles, he had little difficulty fitting them to my sloping roofs. At the end of the installation there were some shingles remaining. The image of the structure on which I installed the leftover material gained the attention of the Copper Development Association audience. My outhouse with a copper shingle roof elicited the morning's longest round of laughter and resulted in its first comic moment. This concluded my presentation. The following speaker, the president of

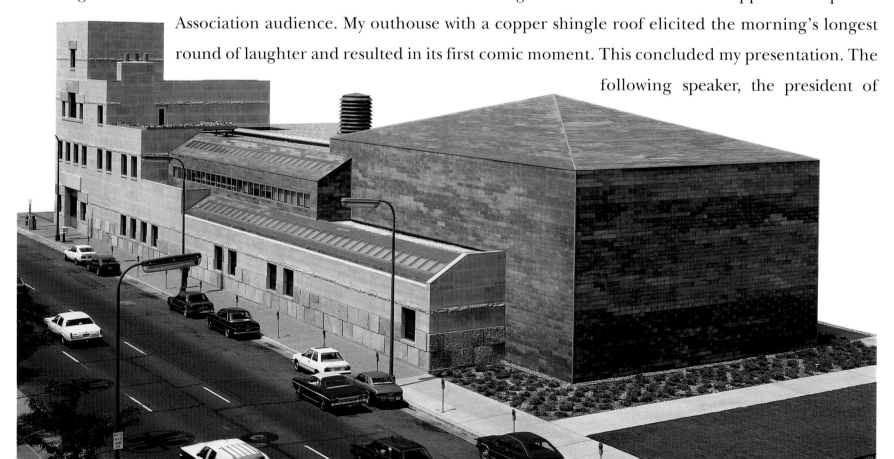

Revere, began his remarks with the second comic moment; he suggested that many in the audience had missed a global marketing opportunity.

My illustrations and comments prompted a number of individuals to come forward with other interesting examples of copper usage, and the invitation to visit the Revere plant was part of this exchange. I accepted it on the spot. A visit to the plant was long overdue, because I had been using copper for years without having observed how it is made. Now I am better able to use the material in a manner consistent with its manufacture, and I am searching for an opportunity to use a portion of the large Revere copper cake in a future project.

With the exception of copper, aluminum, and zinc sheeting, metal products usually play supporting roles in architectural installations. Concealed under finished surfaces, these products are not visible in an occupied building, though some of them have considerable aesthetic appeal. Metal lath, with its open regular striations, is one of them. Prior to 1930, wooden slats on interior wall and ceiling framing acted as lath to adhere plaster. The wood lath's rough surface and the space between the slats worked well for this purpose. In steel high-rise structures tile and concrete were the first choices for fireproofing but metal lath and plaster fireproofing weighing one-half to one-quarter the weight soon supplanted these to reduce the dead load of the building, size of the structure, foundations, and costs. Metal lath supplanted wood because galvanized sheet goods, 2-by-10 feet, could be factory made, incorporating a network of uniform openings and folded ribs for plaster adherence, and delivery to construction sites of compactly bundled packages was economic. Additionally, the installation process was  quick and economical. By the 1970s, gypsum wallboard and metal studs supplanted lath and plaster for similar reasons: ease of manufacture, shorter installation time, and reduced expense. Metal lath did not disappear; it continues to play a role in historic restoration and in installations where plaster is preferred. It also found a new life as an underlayment on the exterior of buildings with plaster and synthetic finishes. I find metal lath to be visually arresting; it is also inexpensive.

Locating a material to meet the functional demands of a modern theater proscenium arch was the challenge in the design of performing arts centers in both Eugene, Oregon, and Anchorage, Alaska. In a traditional proscenium

Opposite: Copper shingles, WCCO Television Communication Center, Minneapolis, MN.; *above:* Copper shingle outhouse roof.

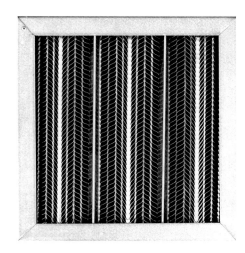

auditorium every audience member looks through this frame to see the theatrical presentation. For centuries, proscenium arches of theaters, opera houses, and musical halls had been made of plaster, which was highly decorated and frequently embellished with paint, murals, and gold leaf. By 1980, the technology of theatrical productions had caused changes in theater proscenium areas. New electronic production elements enhanced presentations, but the need for sound-reinforcement equipment, speakers, electronic gear, and wiring was growing exponentially. These elements could not function behind the plaster proscenium, and, incompatible with the elegance of most auditoriums, could not be placed on the audience side. During the design work on the multi-halled Hult Center for the Performing Arts in Eugene, the acoustical consultant Chris Jaffe and I debated methods of installing this gear. The acoustician formally questioned the use of plaster, an outdated material. The impetus to locate an acoustically transparent surface, to allow equipment placement behind the proscenium, was the culmination of this interchange.

This discussion called into question a traditional element of theater design. It eventually resulted in my suggesting expanded metal lath; its variegated surface could be acoustically open and, with the proper lighting, visually opaque. Jaffe considered my proposal. He set in motion an evaluation of the openings in the lath to determine whether they were large enough to be acoustically transparent. It proved acceptable, and now a radial arrangement of rib lath panels surrounds the proscenium arch in the Hult Center's Soreng Auditorium. The patterning of the sheets was important, because they framed the presentations. Behind this surface, invisible to the audience, is the technical equipment, which is easily accessible from the stage.

Above: Sample, metal lath.
Right: Metal lath proscenium surround, Soreng Hall, Hult Center for the Performing Arts, Eugene, OR.
Opposite: Metal lath proscenium surround, Atwood Concert Hall, Alaska Center for the Performing Arts, Anchorage, AK.

After this success, a more decorative approach developed in the Atwood Auditorium, in the Alaska Center for the Performing Arts. The angled lath panels intersect diagonally along a common meeting line, making a distinct herringbone pattern of raised ribs and mesh. Meeting the acoustical and technical demands of the proscenium arch defined the design challenge and the selection of a suitable material.

These favorable results led me to deliberate about the use of metal rib lath in other applications. One of the most monotonous surfaces in contemporary buildings is the hung ceiling, suspended from a roof or floor structure to provide a finished surface in a room below. Normally, acoustical tiles set in a grid of metal runners forms a flat ceiling with lighting and mechanical systems above. This type of assembly is easy to install and reasonable in cost. The success of metal lath in the concert halls gave me the confidence to contemplate its use and to ultimately employ it in suspended ceilings. Long sheets of lath cut to the size of ceiling panels, with sound-absorbing material above, and integrated with lighting and mechanical systems, have become a new convention for my office. On their visible side,

the ribs in the lath provide direction to the material. Pieces installed in alternating order, turned 90 degrees to each other as seen in the installation at the Student Union Building at Texas Tech University, generate a checkerboard pattern, and running in the same direction, as they do in the ceiling at ImaginOn, they accentuate the raised ⅝-inch linear ribs. This material generates large-scale patterns that fibrous tile ceilings cannot match.

For more than 100 years, sheet-metal building products have been used as decorative surfaces. Because of this sustained application, the public and architects have grown accustomed to seeing them only in certain circumstances. Zinc-coated steel roofing can be seen in many parts of the country; it is inexpensive, fabricated in sheets, readily available, and easy to install. The sheets can be stamped with patterns—my favorites simulate shingles. This smaller design provides a change in scale that is often suitable for my building projects. The tower roofs of the Weber Fine Arts Building at the University of Nebraska, Omaha, the entry soffits at the Mary D. and F. Howard Walsh Center for Performing Arts at Texas Christian University in Fort Worth, and the roof and walls of the Highland House, in Madison, Wisconsin, all display this material.

Above: Metal lath proscenium, Atwood Concert Hall.
Opposite: Stainless steel panels, stage house extension, Kansas City Music Hall, Kansas City, MO.

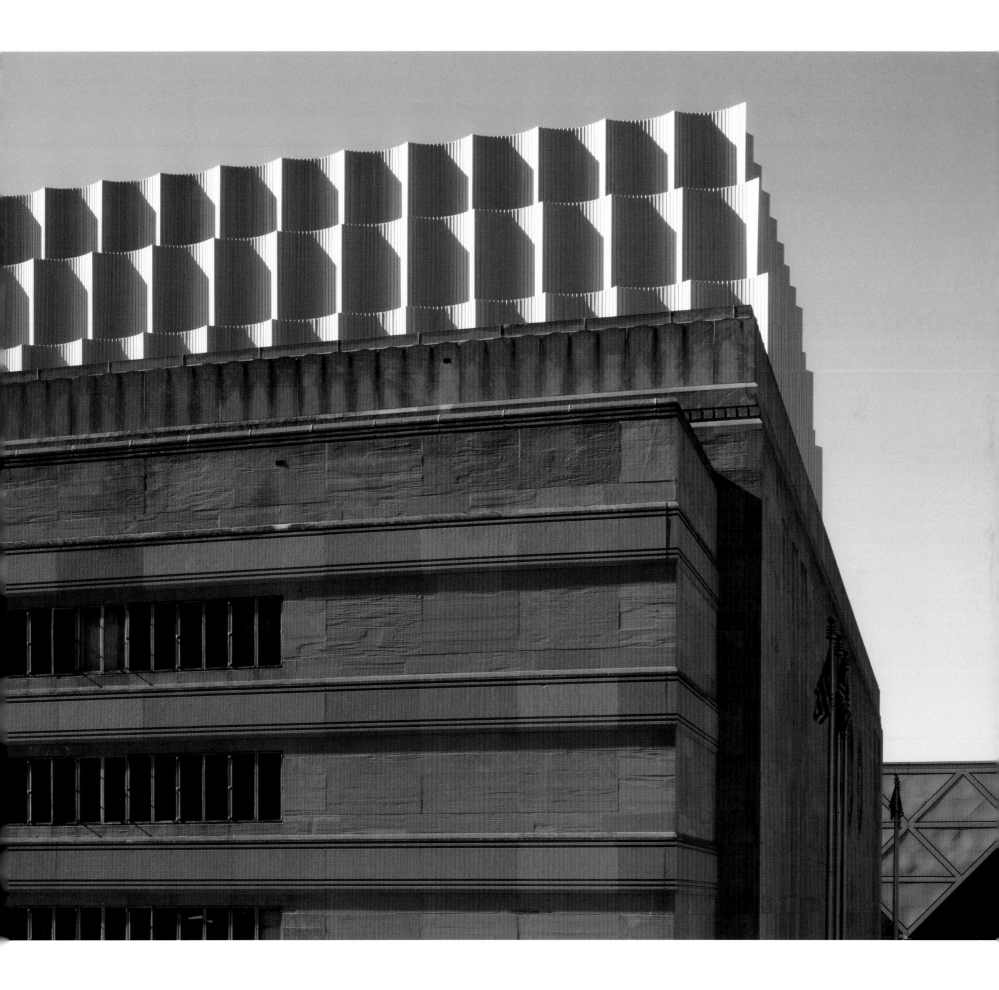

In the early 1900s the fashion to make wood buildings appear more substantial than they were resulted in placing metal sheets stamped with brick and rectangular stone patterns over many structures. When carefully erected, painted, and viewed from a distance, they appeared to be masonry. Only upon close inspection are the overlapping sheets of metal and nails visible. These 2-by-5-foot sheets of galvanized metal are inexpensive, quickly erected, lightweight, and lend themselves to creative installations to this day. During the last 20 years, this pattern has found a new application: the affordable, prefabricated metal home. Upon arrival at their site, before these mobile homes are set onto a proper foundation, the wheels that carried them down the highway are removed. The space previously occupied by the wheels remains, leaving a gap between the home and the ground. One popular way to fill this gap is to cover it with patterned metal sheets that look like masonry. The metal-to-metal construction is compatible, but I believe the implied solidity of masonry, even if it is metal, recommends its use to the new homeowner.

For many years I have utilized the stamped metal products offered by W.F. Norman Company of Nevada, Missouri, initially presented in a catalogue and now found online. In these illustrated compendiums, almost all the sheeting types available 100 years ago are still for sale. They range from ceilings for commercial spaces to exterior siding. In 1987, when my family and I moved to a loft, this catalogue came to mind. Defining each floor in the century-old building was a standard structural system for the period, a center row of columns and matching pilasters projecting from the side walls. The projecting pilasters were an inconvenience, because, with windows between them, they allowed almost no space for hanging objects. It occurred to me that I could bring the wall forward to align with the pilasters except for the areas around the windows. When the plaster was removed from the pilaster, I discovered the fireproofing material was brick. I decided to leave the old brick exposed. A natural choice for the new adjacent surface, therefore, was matching masonry. I turned to the W. F. Norman catalogue and chose masonry-patterned metal wall siding. With the installation of new panels, we now have enough wall space to hang many large objects. Equally important, we delight in seeing old bricks next to new galvanized bricks and stone blocks made in sheet metal from 19th-century stamping equipment. The installation of the two patterns, pressed brick and pitched-faced stone in alternating bands, accentuates the difference in their sizes and allows light to glance off their rusticated surfaces, making them seem more three-dimensional than the real brick.

Using perforated metal sheeting with regularly spaced openings for a variety of architectural applications is an old custom. It is readily available, and easy to customize and install. For the David Saul Smith Union at Bowdoin College, in Brunswick, Maine (pages 152–153), we exploited the versatility of the material in three installations in one space.

Opposite left: Metal lath ceiling, Student Union Building, Texas Tech University, Lubbock, TX.; *right:* Metal "masonry," mobile home installation.
Above: Alternating metal "masonry" panels, Landsman Holzman Residence, New York, NY.

The Hyde Cage, designed by Allen and Collens in 1912, was a single large volume defined by a clay floor, suspended running track, and rooftop monitor. Converting this structure to a new use included preserving a sense of the volume of the space and the monitor. Initially the electrical engineer for the project suggested the use of high intensity light fixtures to augment the natural light from the monitor in the central space. These compact fixtures have grown smaller as they have become more efficient. A problem of scale existed in using them as suggested; these small objects floating down from the roof surface would look tiny in this 40-foot-tall interior space. To remedy this, a circular perforated metal corona designed for attachment to the basic catalogue unit enlarges its presence in the room. Additional light was required along the edge of the gently sloping ramp that rises through the space. A series of inexpensive special lighting fixtures consisting of floor-mounted poles, terminating at the top with two fluorescent bulbs sheathed in perforated metal with a serrated semi-circular outer edge resembling the college seal shape, provide navigation points along this open walkway. One large clock uses this same serrated edge to tie all three elements together in the space and make them unique to the college. During the day, modulated light permeates the space from a continuous band of windows in the monitor. The three additional sources provide complementary illumination at other times of the day.

Manufactured in varying profiles and in a range of depths, standard aluminum extrusions provide hard, long-lasting surfaces that require minimal maintenance. Frequently found at storefronts in shopping malls in 4-inch or smaller widths, these 20-foot-long, ¼-inch-thick pieces seemed ideal for other applications. Inspired by their durable surface, several standard shapes not usually assembled together form a larger pattern to animate and minimize wear on the sidewalls of the escalator runs in the heavily trafficked atrium of the Tom Bradley Wing of the Los Angles Public Library.

A similar product application, intended for one use but applied in another, is the industrial stair tread. These steel units, pre-manufactured with perforations, normally allow fluids to drain away and reduce the total weight of the stair. Large openings and raised buttons provide good footing for use as stair treads, ramps, walkways, and loading dock floors. They come in standard-sized lengths starting at 2 feet and in 6-inch increments up to 4 feet. Turned upright, painted, and with an appropriate connections, these units make fine decorative supports for handrails at a modest cost. Although made of sheet steel they have a large proportion of their surface area open, making the railings secure while providing a sense of airiness. Installed as decorative railings in highly trafficked and public areas at the Columbia Public Library in Missouri, and the George A. Purefoy Municipal Center in Texas, the users of these spaces seldom know their original intended use. At the Courtyard Theater in Plano, Texas, they form a continuous balcony fascia; the pattern of perforations in each panel provides one of the few ornamental surfaces in this 250-seat, flexible community auditorium.

Opposite left: Metal roof shingles, Weber Fine Arts Center, University of Nebraska at Omaha, Omaha, NE.; *opposite right:* Perforated metal light fixture, David Saul Smith Union, Bowdoin College, Brunswick, ME.; *above:* Metal stair tread railings, George A. Purefoy Municipal Center, Frisco, TX.

Following the completion of the 1970 campus master plan for Shaw University in Raleigh, North Carolina, one small project was implemented. The master plan introduced a college of arts to the curriculum, which expanded and integrated academic and community activities. The Community Service Center located on a corner of the campus adjacent to the surrounding neighborhood brought together several college departments for seminars, workshops, community meetings, and student activities.

The economy of prefabricated metal buildings is well known. These precut manufactured structures, seen in all parts of the country, are easy to erect in a short time, are readily obtainable, and are inexpensive in comparison to commissioned structures. At the time the campus plan was completed the university could afford little more than prefabricated buildings. Two ready-made structures were brought together to form a breezeway, distinguishing them from similar structures. Omitting one corner section from one structure allows a diagonal orientation for the breezeway between. This simple gesture and the interior partitioning are the only customized parts of the standard units. The facility is still in use today.

In 1972, at Exeter Academy in Exeter, New Hampshire, when Louis Kahn, one of the most influential architects of the mid-20th century was bringing an alternative modernist architecture to this 200-year-old campus, we were devising a way to build a theater under severe budgetary constraints. The only affordable scenario for the project relied on a prefabricated building enclosure. The economy of the structure, roof, and exterior walls allowed for an offsetting of the standard rectangular volume to shape a special interior space. A focused main seating configuration, technical grid, additional movable seating, and other theatrical elements occupy the main volume of the structure. A lower level provides support and minimal lobby space. It, too, is still in use as a teaching space.

In the 1990s, two decades after these exercises in economy, I returned to prefabricated buildings when confronted with limited budgets for two projects: an addition to the Salisbury School in Salisbury, Maryland, and the Fox Theater, a multiplex cinema in Wyomissing, Pennsylvania. The acoustical and technical requirements for the theater proved too

Above: Prefabricated buildings model, Shaw University, Raleigh, NC.; *opposite:* Metal stair tread railings, Courtyard Theater, Plano, TX.
Following pages: Salisbury Upper School, Salisbury, MD.

stringent for a structure that was entirely prefabricated, but investigation suggested a hybrid. The Wonder Trussless, an arched building system similar to a Quonset hut, gains strength from bent and crimped steel sheets bolted together. This system spans large distances without needing columns for support; 16-gauge crimped sheets form corrugations in one direction and are bent in the other direction to form a rigid structure. At

the Fox Theater, these metal panels clad the exterior walls of the auditoriums. Individual crimped sheets run from ground level to the roof, accentuating the deep corrugations. At the Salisbury Upper School, these panels enclose arched passageways from the center of the building to its radiating wings and the main entry.

The ability to use materials economically results from my knowledge of the process of production, the physical properties of the product, and the product's limitations. This information enables me to utilize materials in unusual contexts and

to install them in novel ways. There is a cost to the architect in utilizing materials in this manner; it takes time and initiative to find a material, investigate its possible application, develop construction details, and supervise its installation. However, undertaking these fundamental parts of the research and design process is gratifying; it allows the physical presence of a building to appear rich and appealing while its materials meet limited budgets.

Opposite: Extruded metal escalator base, Tom Bradley Wing, Los Angeles Public Library: Central Library, Los Angeles, CA.; *above:* Metal passageway, Salisbury Upper School. *Left:* Metal shingles and cresting, George A. Purefoy Municipal Center. *Following pages:* Metal siding, Fox Theaters, Wyomissing, PA.

06 | Against the grain

Timber is now processed for construction in much the same way as in prior eras, but today it occurs with the aid of computers, lasers, and sophisticated machinery. It no longer requires brute strength and workers using hand implements to make building products. The conversion of raw lumber to finished wood products can now occur in less than an hour.

On a hot, humid, summer morning in 2005, I visited the Cheat River Valley on the outskirts of Kingwood, West Virginia, to tour the Allegheny Wood Products mill. A series of sprawling sheds, piles of logs stretching hundreds of yards, pallets of finished planks, and moving equipment surrounded the hardwood fabrication plant. This scene was not surprising but the process inside the sheds was. An automated ballet of logs transported on a seemingly endless conveyor under a vast metal roof was a stunning sight.

My tour began when the sawmill manager directed me toward the area of outdoor operations where open-sided trailer trucks filled with various diameter logs arrived at irregular intervals. After unloading, each log was inspected and graded; this process determines the amount of timber delivered by each truck. Extra large logs went in one direction and all others in the opposite direction and into the mill. The first step in the manufacturing process begins when a pincer-like device grips and moves a log while stripping the bark. At the next location in the automated fabrication operations, a computer determines the maximum amount of finished wood each log can produce. A visual and numeric data display

conveys this information to a machine operator who sends the log on its way to become a series of 1-inch-thick boards. A band saw removes the outside pieces from the log as it is rotated for each cut. The remaining almost rectangular portion of the log moves to the gang saw. The result of this cutting process is a series of 1-inch-thick planks with varying edge conditions known as "flitches." To make each flitch rectangular, the board moves through two final steps: a laser-guided operation aligns each board to maximize its width and the final length is set with end cuts. A grading process puts similar size pieces of wood together for placement in palletized stacks for shipping. By-products from this continuous operation—wood chips, sawdust, and bark—become oriented strand board, fuel pellets, and mulch at other nearby manufacturers. In this manner, the entire original log is used.

The design vision for the Center for Contemporary Arts at Shepherd University in Shepherdstown, West Virginia, includes using regional construction products for the primary building components. This concept led to a meeting with the president of Allegheny Wood Products. He was interested in my desire to display the natural character of wood in the

Opposite: Timber truck fully loaded.
Above, left to right: Timber fabrication, from log to board.

arts facility. For the past 50 years, most purchasers of wood products sought lumber without blemish and distinctive grain patterns. My interest in using poplar as the wood for one of the theaters allowed him to know that I appreciated all the qualities that wood products can exhibit. Poplar has irregularities and color variations; sections of green and even purple appear randomly in the wood, reflecting the intake of minerals during the growth process. For my visit to the Kingwood Mill, poplar was run through the fabrication line so I could see the widest range in color variation available. When I departed from the mill tour, two small flitches accompanied me. A day later, the building committee for the Contemporary Arts Center admired these two samples and endorsed their use as cladding for the theater spaces.

The ability to use so ubiquitous a product as wood in a special way has to be earned. The architect must rise above its familiar applications, local customs, and standard details to expose the elegance of its grain, texture, and color. Turning wood into lumber is a common process, but turning ordinary lumber into an alluring material is a far more complicated exercise. I have employed simply milled wood on numerous occasions. For example, the five sloping fir columns that support the raised roof of Temple Israel, in Dayton, Ohio, define the space and add a sense of warmth to the sanctuary, the heart of the building. The pinwheel skylight, at the sanctuary's summit, floods the space with

Top: Fir beams and columns, Hult Center for the Performing Arts, Eugene, OR.; *above:* Wood columns, Temple Israel, Dayton, OH.
Opposite: Timber columns, beams and decking, Temple Israel.

Top: Timber structure, main reading room, Columbia Public Library, Columbia, MO.; *above:* Main reading room, Columbia Public Library.
Opposite: Minnesota Orchestra Hall, Minneapolis, MN.

reflected light. Over time, the wood has opened along its grain, revealing its drying process and the material's natural properties.

A comparable installation at the Columbia Public Library in Missouri has five laminated wood columns supporting the mezzanine, the second and third floors, and the roof of the building's entry pavilion. Laminated beams and ties support the roof to form a canopy over the uppermost library reading room. The tent-like quality of the structure supplies a sense of shelter to the users and frames unprecedented views out to the community.

Not surprisingly, wood is the musician's material of choice for a place in which to perform. Some of the most highly prized musical instruments are made of wood, and by extension, many musicians consider a music hall a large instrument, noting "wood is good." Most acoustical experts, too, like wood, because its mass brings sonic warmth to music spaces. This timbre (as opposed to timber) results from low-frequency reflectivity—the greater the mass, the greater the warmth.

In the early 1970s, while developing the design for Orchestra Hall in Minneapolis, Minnesota, orchestra members expressed their preference for wood. Three walls and the floor that enclose the 2,400-seat auditorium have wood surfaces. At that time, white oak was readily available and its grain and light-blond color appealed to this largely Scandinavian community.

In Orchestra Hall, the materials selected provide optimum characteristics for symphonic music. Wood installed according to acoustical criteria—boards of equal thickness (slightly less than an inch), of constant width and length—form the angled auditorium sidewalls. These surfaces distribute sound from the performance platform to the remainder of the space. Its color offsets the pale blue-gray, heavily plastered, decorative cube-encrusted ceiling and rear stage wall. These two materials interlock to join audience and performer in one space.

The design of large civic auditoriums usually comes with budgets matching their communities' and supporters' ambitions, as public projects reflect, and in time come to symbolize their location and performance organizations. Wood is the material requested most often, and funding is normally devoted to securing it. The budgets for academic and small community facilities are not usually as accommodating as they are for larger projects, thus challenging the architect to achieve the same high level of acoustical excellence with less costly materials. At Middlebury College, in Middlebury, Vermont, the entire student body is smaller than Orchestra Hall's seating capacity. The goal was to bring the campus's arts programs under one roof in a new facility that would contain a flexible theater for plays, a music hall, dance performance studios, and an art gallery that would serve both the institution and the region. It was anticipated that this consolidation would enrich individual programs, facilitate interdisciplinary efforts, and nourish creativity.

The 370-seat Middlebury recital hall, built in 1992, is similar in concept to Orchestra Hall but has one-sixth the seating capacity (see Chapter 4). It contains musicians and audience in the same space; no proscenium separates them. The hall is comparable to an elongated octahedron, an eight-sided faceted volume, with the musicians at one end. Four-inch-thick wood decking on steel beams serves as the structure and finished ceiling surface of the hall. The majority of the audience sits at the center of the space occupying one end of a low-walled, cherry wood-clad ellipse set away from the outer surfaces of the room. The musicians sit in the remaining third of the ellipse. Capping the wall surrounding the audience is a specially shaped wood handrail that terminates at the stage in a volute, a three-dimensional fiddlehead; these two elements have come to symbolize the music space, a tactile embodiment within easy reach.

Because of the project's economic limits, the walls of the hall could not be clad with solid wood, so we explored the possibilities of oriented strand board (OSB), a manufactured product that enjoyed great popularity in the 1980s home building market. Produced from sustainable sources, it became a popular successor to plywood. It does not exhibit the stress characteristics associated with veneer material, and it is easy to install. The largest American manufacturer of OSB annually produces 6 billion feet of this material, which forms the sheathing and sub-flooring in new home construction, providing rigidity to the house frame and a stable backing for application of exterior wall surfaces and floors. It consists of various sizes of wood chips, a by-product of the manufacture of dimensional timber, in an adhesive matrix. At Middlebury, our acoustician agreed to its use after verifying the properties and quality of the compositional elements of the material and the method for rigid installation. Specific acoustical requirements dictated the material's thickness, its panel sizes, and the splaying of its installed surfaces.

OSB is not a finished product, nor is it intended to be a visible surface; a moisture-resistant coating, waterproof edge, and stencil surface labels can be easily seen on both sides of standard 4-by-8-foot panels. It requires finishing, sanding, staining, and lacquering in a woodworker's shop to make it presentable in a concert hall. Many individuals seeing OSB for the first time after it has been finished believe it may be stone, as an off-white stain gives the surface a marbleized appearance. Because of its low cost, sustainable source, wide availability, and woodworker comfort in handling it, this music room has an all-wood interior, even if one of its surfaces is a factory-made product not originally intended for high visibility.

The self-assurance to use OSB in a prominent public place began modestly with a private residence in 1985. Marty and Rita Skyler owned a parcel of beachfront land on eastern Long Island. Their budget for construction was limited but they desired a special series of connected living spaces for enjoying their Atlantic Ocean view. Minimal maintenance in future years ruled out the application of standard home products—gypsum wallboard and paint were not acceptable. As an alternative, I suggested OSB. Almost all of the walls forming the open interior spaces are white-stained, 16-by-32-inch

Opposite: Cherry handrail, concert hall, Center for the Arts, Middlebury College, Middlebury, VT.; *above:* Sample, oriented strand board (OSB).

sheets installed in a running bond pattern. This successful installation allowed me to contemplate using the material on other larger projects; woodworkers found the material easy to handle and finish in the shop, and it was economic to install.

Following the success of the public installation at Middlebury, OSB has found its way into a number of other buildings, and has advanced from a background position to take center stage. The concept of a "box within a box" is the basis of the design of the Pepsico Recital Hall for the Walsh Center for Performing Arts at Texas Christian University (TCU). The exterior box provides the acoustical volume and physical enclosure required for the musical performance while the inner

box supplies a comfortable visual space for small ensemble and solo presentations. Large openings in the timber and OSB surfaces connect the two areas making one acoustical volume. A 3-by-1-inch wood strip available in home-improvement stores forms the leading edge of the exposed timbers. Its undulating gold finish reflects the house and stage lighting, adding highlights and sparkle to this intimate music room. Intended as a roofing closure, this edge strip typically fits into the profile of the corrugated contours of sheet metal and plastic panels. Its low cost, availability, ease of installation, and versatility recommend it for other installations and it has found decorative applications in several recent projects.

Opened in 2006, the Carol Bush Emeny Performance Hall at the Globe-News Center for the Performing Arts in Amarillo, Texas, serves as a public performance venue complementing the nearby 2,300-seat auditorium in the convention center.

Above: OSB walls, Skyler House, Bridgehampton, NY.; *opposite:* PepsiCo Recital Hall, The Mary D. and F. Howard Walsh Center for Performing Arts, Texas Christian University, Fort Worth, TX.

The total project budget for this facility—$30.6 million—was sizable for a privately funded community undertaking, but it was modest in comparison to auditoriums built in Texas and other cities across the country at the time.

On first viewing, this auditorium appears similar to the music halls at Middlebury, Orchestra Hall, and TCU, where one room contains both audience and performers. A wooden, three-part, basket-like surface that extends from the walls and across the ceiling surrounds the audience and performers. The 1,300-seat multipurpose auditorium looks like a music hall with highly articulated surfaces defining the room. However, with the orchestra-shell section of the room enclosure removed to a rear stage position, it also functions as a proscenium auditorium with a full stage tower and rigging. A variety of performances utilizes the full stage, wing space, and fly loft. Even though the basket-like stage portion of the structure weighs almost 30 tons, it moves mechanically by overhead crane to its offstage location in less than four minutes.

The curvaceous wooden basket spanning the audience–performer area consists of 12,000

Opposite: PepsiCo Recital Hall.
Right: Timber roof, library reading room, George A. Purefoy Municipal Center, Frisco, TX.

linear feet of 3-by-12-inch laminated timbers varying in length up to 12 feet long. They act as a structural armature, with small steel connectors and bolts tying them together at their intersections. Filling the rhomboid spaces between the timbers is 75,000 square feet of stained OSB. Together the OSB and laminated timber form the room's acoustical and finished architectural surfaces. This wood enclosure sits within a larger masonry enclosure. Both the space within the basket—the audience and performer area—and the space between the basket and the outer building serve as the reverberant acoustical chamber for the auditorium. A warm stain applied to the wood surfaces gives the basket a sense of depth, while lighting animates the shell's surfaces. Although every piece of wood and OSB is flat and rectangular, they combine to form the large, seemingly curvaceous surfaces that envelope the audience and focuses them on the stage presentation.

An enquiry from a previous civic client set me on a course to modify an existing kitchen in a recently acquired "classic six" pre-war Manhattan apartment. Upon inspection, I discovered a mean galley kitchen: white plastic laminate counters, oversized floor-to-ceiling cabinets, and equipment down two sides of a 6-foot-9-inch-wide, 12-foot-long room. The only element of interest was a small window with a southwest view into a light well at the rear of the building, facing other backyards.

The cook in the family is a small individual, barely 5 feet tall. By comparison, her husband of three decades seems almost twice her size. Upon a few days' reflection, the volume of this tiny space seemed to me to suit her diminutive stature. They were both delighted and surprised

Above: Movable music shell, Globe-News Center for the Performing Arts, Amarillo, TX.; *opposite:* Globe-News Center for the Performing Arts.

when I suggested that they might eat in their new kitchen as well as cook, since they could barely fit in the current one simultaneously. I decided size was a virtue and exploiting it was the only option for success.

A galley-type format needed to remain to accommodate all their desired cooking equipment, china, and glassware. To break the tyranny of the parallel walls and cabinets of a galley, I decided to make an implied center to the space. A series of eight overlapping layers of curving horizontal cabinets line some part of every wall. The illusion of a single centralized space draws the eye around the room to the source of natural light where the millwork is absent and then back to the center.

To build this fanciful series of cabinets required a special woodworker. For more than two decades I have worked with Oscar Hoff III, a Detroit artisan, to make special installations. By working in his studio, fabricating dimensionally intricate millwork is easier and economic to achieve. Eighth-inch bendable plywood with Baltic birch veneer formed the finished cabinetry surfaces, some with sliding doors and others with green opaque glass. Ebonized walnut selected for its dark color graces the countertops. To increase counter space, under-the-counter refrigerator and freezers were selected, a compromise for the husband but excellent for the wife. One movable counter nests in front of an access door, during meal preparation and at dinner service it occupies the kitchen entry. For the last seven years, the clients have not eaten a meal at their dining room table in an adjoining space except for the evenings when they entertain guests.

Selecting wood veneers to generate a sense of warmth and comfort in a minimal space was the easiest part of this endeavor. The fabrication of the cabinetry was the critical aspect of the undertaking. Oscar's careful attention to each section of the installation made it all come together as a whole. Since completion, the husband has confided in me that the movable counter section is a beautiful piece of sculpture. He went so far as to anticipate some future day when they might be forced to move; if this occurs he will consider taking this part of the kitchen to a new place.

Wood is an instantly recognizable material yet varies vastly in appearance. Appreciating its complexities and linking them to a design may begin with a flickering insight. First glimpses of recognition and clarity from pondering an ambiguous situation and finally teasing out the subtleties and distinctions of a specific wood species can provide memorable architectural results.

Opposite: Eight cabinet layers, Holman kitchen, New York, NY.

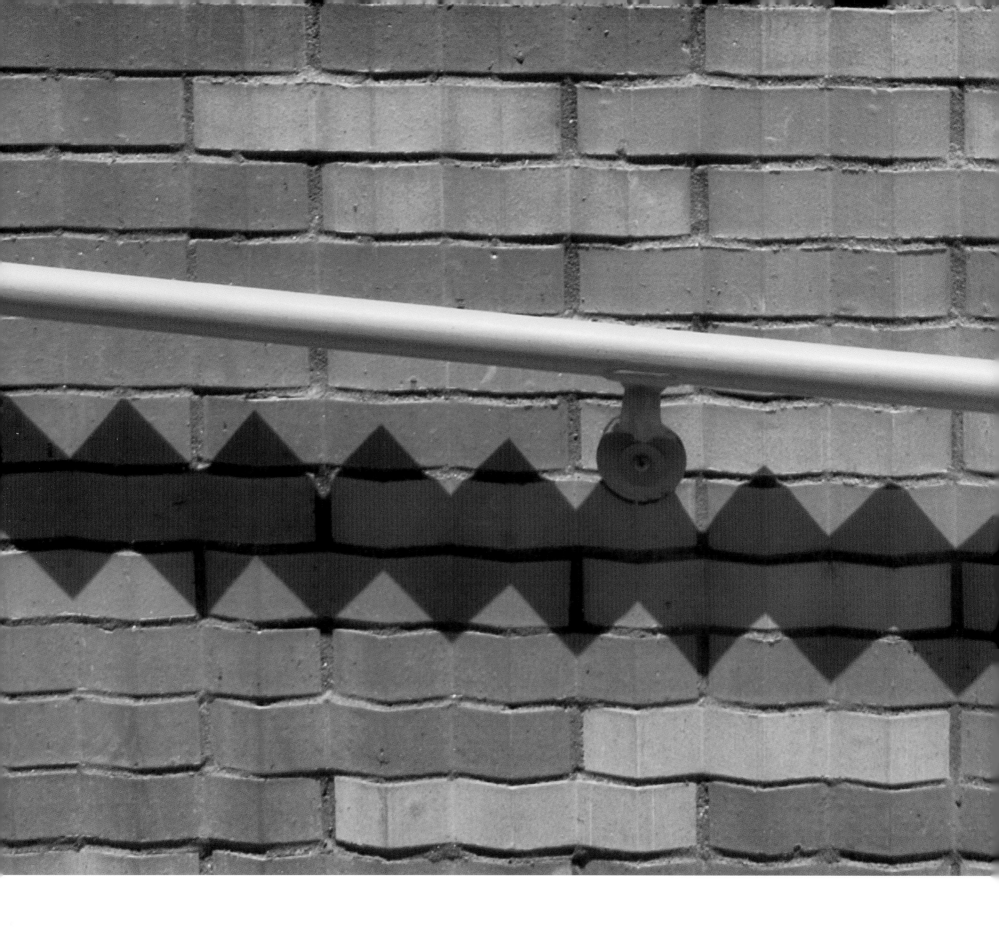

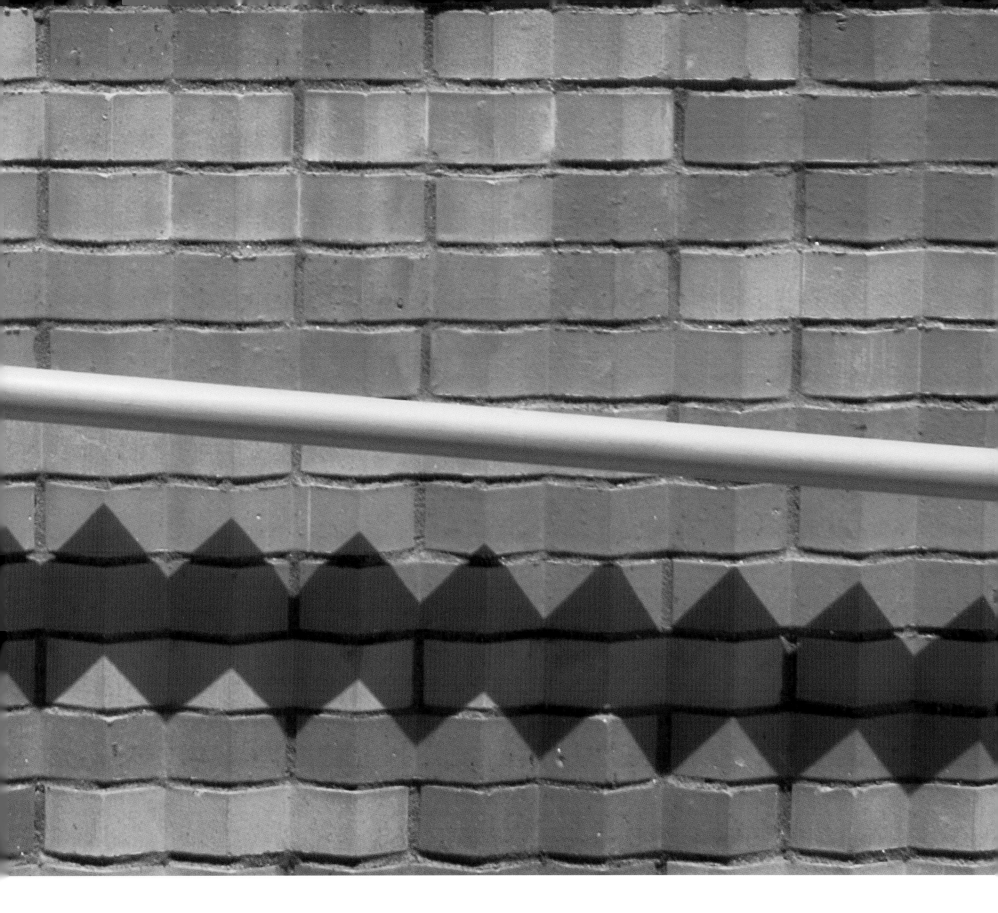

07 | Shaping clay

A building's program—its overall function and its spaces' specific uses—is one major component of what distinguishes its built outcome from other structures. Preparing the program statement is an initial project undertaking for architects. I listen carefully to the client's stated needs and focus on distinctive elements as they arise. These idiosyncratic program elements are the key to developing truly singular designs. This early pre-design work establishes more than the client's needs: the site location, project cost, construction schedule, and materials are often determined. People from numerous spheres help define the building's program. Civic undertakings may include users, members of the community, government officials, and donors. At academic institutions, the group often consists of faculty, administration, students, governing boards, and benefactors.

Some architectural firms have repeat clients, who commission similar buildings in a brief three- to four-year period. This repetition usually occurs in commercial, retail, and residential development, which may not require extensive program or design work for each undertaking, and which replicates aspects of previous projects. Repeat clients in the arts, however, are infrequent, because communities don't usually construct similar public buildings within a limited time frame. Because the opportunity to build a performing- or visual-arts facility arises so rarely, many clients for these projects are not familiar with the programming process that balances ambitions with budgets and other concerns. More time than anticipated is often spent aligning these client goals.

In the late 1980s, I had the pleasure of developing arts programs for two academic institutions at the same time, both of them in Omaha, Nebraska. Though they were not exactly repeat clients, one of the institutions benefited from this simultaneity. Each project mandated a significant performance space to serve its university community and the surrounding city. The completed programs reflected the differences between the institutions and their constituencies, as did the resulting buildings and the materials selected for construction.

A faculty committee of about a dozen individuals, who represented the disciplines that would use the completed structure, developed the program for the Lied Education Center for the Arts at Creighton University, adjacent to the downtown business district. A variety of arts programs, from photography to music to ceramics, made their needs known through open meetings and workshops. The chair of the committee, Jerry Horning, a ceramicist, organized them, served as liaison to interested parties, and oversaw completion of the program work, on schedule as it turned out. Jerry did much more than secure a timely completion to the program portion of the project; he orchestrated the participants about the intangible qualities of the building, the amount of academic space shared by departments, and the ways in which the arts programs would be presented to the university community. Equally important, he introduced me to the source of a local building material.

During a discussion of materials for the new arts building, Jerry described the Omaha Brick Works Inc., a small, family-run business. It was a modest operation, not widely known even in Omaha, as it did little in the way of self-promotion. Jerry's curiosity and wide-ranging interest in clay had originally taken him to this brickyard. The proprietor, Maury Cullen, had taken over an older operation that dated from the beginning of the 20th century. It included one rectangular and three beehive kilns to fire material. From time to time Jerry, other faculty, local artists, and students participated in special projects using these century-old kilns. An occasional student would work for Maury.

My first visit to the Omaha Brick Works occurred spontaneously during a break in one of our programming days at the university. We had 90 minutes between sessions for lunch, and Jerry inquired about my interest in a visit to this brickyard. My immediate positive response resulted in a quick telephone call, to make sure Maury was available, and then a 15-minute drive took us to Ralston, at the edge of the Omaha city limits. The brickyard was literally on the wrong side of the tracks; access

to it was under a railroad trestle. Later, I would learn that this single-lane route, between supports of the overhead tracks, severely limited access to the plant by delivery trucks.

The company's beehive kilns were about 30 feet in diameter. With their walls, almost 3 feet thick, and an oculus in the ceiling, they looked more like the Pantheon than a furnace for firing 60,000 bricks at one time. The space inside the kilns was dramatic; each dome's inner brick surface was black and showed years

Left: Beehive kilns, Omaha Brick Works, Inc., Omaha, NE.

of use. This was the case except where the heat from firing the clay was so intense that the surface had failed, revealing raw brick. Light filtered in through the oculus. The only evidence of maintenance was the occasional replacement of interior brick and plastering on the exterior of a few of the domes. The only modern improvement was the introduction of gas lines that supplied heat for the firing process.

On this first visit, I saw raw material being prepared for firing, kilns in operation, finished bricks, and a few student art projects. To one side of the yard was a spoil pile containing discarded material; this area, of course, immediately attracted my attention. From my conversation with Maury, I learned that the heat in the kilns was difficult to control, and when bricks weren't properly stacked in the kilns, "clinkers" resulted. Occasionally, parts of adjoining bricks fused during firing, one brick would adhere to a section of another, leaving black burn marks. The bodies of these bricks appeared normal, but the finished faces were deformed; no two had the same appearance. I left Omaha Brick that day with a few clinkers weighing down my bags. I didn't know where the bricks would be used, but I thought they had potential given the proper circumstances. As we returned to our program assessment meetings at Creighton, I thought about the brick clinkers.

As work proceeded at Creighton and further discussions occurred about materials for the project, it became apparent that a light-colored building, similar to a nearby, main-campus limestone structure, would be appropriate for its new arts facility. This ruled out the use of Maury's red bricks.

Several months later, Larry Jacobson at The Schemmer Associates, my local associate architect for both projects, invited me to the parking lot outside his office to review brick samples. He knew it was time to consider materials for my other Omaha project, the Weber Fine Arts Building at the University of Nebraska, 40 blocks further removed from the city business district. Larry also invited a local representative of a national brick manufacturer, which had supplied materials for other campus building projects. The presentation was outdoors, not only because the rep didn't want to drag the

Above: Sample, clinker bricks.; *opposite:* Alternating clinker brick bands, Weber Fine Arts Center, University of Nebraska at Omaha, Omaha, NE.

materials up three flights to the office, but also because color variation is best judged in natural light. Arranged between parked cars and the sidewalk were panels displaying ten different types of red brick. Each panel contained eight brick faces to provide a sense of the color range and finish for each type.

The representative made a long introduction about quality control at the manufacturing plant—the production of bricks had become a "science." Strict controls guaranteed that every brick type would be consistent, variation between individual bricks of the same type would be minimal, and imperfections would be impossible to detect. Uniformity was the goal, and they could achieve it. The brick samples he chose to show me matched other university projects and illustrated his remarks. This presentation immediately brought to mind Omaha Brick's clinkers—the contrast could not have been greater. Compared to the machine-made samples in the parking lot, Maury's bricks looked irregular and carried the imprint of the manufacturing process.

This sales presentation set me on the opposite course from the one Larry anticipated. After this meeting, I examined Maury's bricks even more closely—their structural qualities, initial rate of water absorption, and the amount of irregularity that might be acceptable—while weighing the effect of the irregular shape against the additional time required to install the brick. From the start, the Omaha Brick clinkers were exceptional, far more interesting to me than the unvarying product presented by the sales representative. I realized that Maury's irregular bricks would soon be a thing of the past if the sales representative was right about the production and demand for uniform building products.

The building conceived for the University of Nebraska was an assemblage. The largest program elements—theater, art gallery, elevator, and stairs—occupied their own volumes, set apart from the small spaces—studios, offices, and labs—contained in a single central element. The small spaces represented the majority of the building, with public access at the center of a line of these rooms arranged in a linear fashion. This elongated three-story bar was ideal for an economic application of brick. After validating the quality of Maury's bricks, a strategy for evenly mixing clinkers with standard shapes of the same color evolved. At the time of the masonry mock up review, we abandoned this approach since the masons asserted that this mix of bricks required greater installation time than originally anticipated. Instead, a series of alternating bands allowed the masons to install one type of brick at a time over the 35-foot height of the wall without a change in the project schedule. Larger clay units stacked vertically formed a base for the brick bands. The irregularities in the bricks, accentuated by the gently undulating 240-foot length of wall, picked up sunlight as it moved from the main campus walkway to the parking lot at the opposite end of the building.

Maury's brick business was more a craft operation than an efficient industrial process. Jerry Hornung understood that I could grasp the potential of these bricks only by visiting the plant. It was ironic that Jerry introduced me to this special brick operation because the material appeared at my other project at the University of Nebraska, Omaha. Shortly after my last project with Omaha Brick, Maury's son informed me that the operation would close. This news was disappointing, but not entirely surprising, as many small family-run operations are unable to compete with larger enterprises. I have not gone in search of clinkers lately, but I assume they will be difficult to locate. Though the demise of Maury's business resulted from specific conditions, limited access to the brickyard, scale of operations, and competition place many small independent companies in similar jeopardy.

Opposite: Weber Fine Arts Center.

The banks of the Seco Creek in D'Hanis, Texas, are a rich source of raw material for brick and tile production. For generations this town, 50 miles west of San Antonio, has drawn on its surrounding population of 300, as a work force for making clay products. The D'Hanis Brick and Tile Company has been in continuous operation since 1905. Today, John Oberman runs this last remaining manufacturing facility in this small central-Texas community. Raw clay mined from the nearby creek banks is ground and screened, extruded, dried, fired, and packaged for shipment on a weekly basis.

The D'Hanis plant is similar to the Omaha Brick operation. It uses century-old beehive kilns to fire clay, it operates under independent ownership, and it has been in operation for more than 100 years. Its quality control, its variety of shapes and textures, and its blended range of color products distinguish it. Non-automated operations like D'Hanis allow architects to produce limited runs of brick and tile not easily made by larger companies for little difference in the cost of a finished unit. For these companies producing a die for a new extrusion is a challenge rather than an expense.

D'Hanis red-clay brick and tile play a major role in several Texas projects my firm has designed and constructed in the last decade: the Walsh Center at TCU, the Murchison Center at the University of North Texas, and the Performing Arts Center at Texas A&M University-Corpus Christi. The material's distinctive red color, well known to regional architects, can be found in many buildings across the state, rooting structures in the quickly changing Texas landscape. Special sizes, shapes, and textures, including fluted, scored, zigzag, and scratched surfaces, grace the walls of many of my Lone Star projects.

Palletized D'Hanis tile units travel easily from their native state to distant locations. At Temple Israel, along the Miami River in Dayton, Ohio, clay units from D'Hanis play a role in shaping two of the project's most striking features. The diffused overhead natural light on the 5-by-8-by-12-inch scratched and smooth-face structural tile gives the main sanctuary an earthen color, and forms the background to the bima and ark. The sanctuary and the building's front, curving façade are composed of standard red-clay shapes, including 2-by-4-foot ribbed rectangular sections, octagonal units (normally used as drain

tiles), quoins, and coping from the Superior Clay Corporation in Uhrichsville, Ohio, and rectangular units and solar blocks, cut in half to expose their cylindrical forms, from D'Hanis. The tiles are set to accentuate the front wall's curvature and length in southern daylight. The surfaces of the sanctuary and the façade, although remote from each other physically, visually connect the building exterior and interior in this contemplative riverside park environment.

Though brick operations are mostly automated these days, when visiting them I still search for distinctive products. Even complex mechanized operations fortuitously produce discoveries. In Elgin, Texas, looking for material for the Walsh Center at TCU, I visited the Elgin Butler and Acme Brick plants. Acme is the largest American-owned brick manufacturer. Finding a color match for the prevailing campus brick, a blend of yellow and beige with a scratched face was the objective. A light yellow Elgin clay tile appeared similar in color to the TCU blend. After discussing its potential use, we visited the Acme operations across the road. During this part of the tour, I remembered a previous visit to another brick operation

Opposite: D'Hanis tile and Acme brick frogs, Mary D. and F. Howard Walsh Center for Performing Arts, Texas Christian University, Fort Worth, TX.
Above: Performing Arts Center, Texas A&M University-Corpus Christi, TX.

because I saw row upon row of stacked bricks on pallets showing their rear faces. These surfaces displayed "frogs," the regularly spaced reveals in the clay resulting from the negative impression of the guides from the extrusion process that forms the brick units. The frogs' dimensions were about the same in width—⅜ inch—as the typical space left between bricks for the application of mortar.

The series of repeated reveals presented the visual opportunity to change the scale of the brick and the wall surface from the dimension of the brick unit to the dimension between frogs. This change in appearance could occur at no additional cost in the manufacture or installation of the brick. I requested Acme to produce a sample brick with one significant change: frogs on the finished face of a typical TCU unit as opposed to the rear face. This request, for the frogs and scratch finish to occur on the same face instead of opposite sides, was puzzling to the manufacturer, but it allowed for the potential change in scale. The finished brick sample, produced from an Acme plant in Arkansas, later met with design and building committee approval. The frogs on the bricks' faces changed the perception of the finished brick installation. The largely windowless walls surrounding the music rehearsal spaces and shop area at the Walsh Center appear as large surfaces rather than a series of small bricks. Averting the monotony of a standard brick installation so impressed TCU that the next university project, a technology center, also used brick with the frogs exposed.

Brick has been an indispensable building product since ancient times. Every user of the material shapes the clay to their special needs. The inhabitants of Pompeii erected the most imposing public structure in 120 B.C., the Basilica, to serve as a law court, business center, and meeting place. Large columns form side aisles down the length of the rectangular structure. A long tradition in using this building form would evolve into the model for early Christian churches. Equally interesting are the circular and elongated pentagonal brick shapes used to form the understructure of the 28 columns that shape this space. The bricks, originally  stucco-covered, can still be seen at this ruin of an historic site. It is evident that the two brick types allowed for ease of manufacture and installation of these monumental columns. This long tradition of shaping bricks for specific purposes carries forward to this day.

During the material selection process for the Weber Fine Arts Building, I became better acquainted with the Omaha Brick operation. One of the standard products, Roman brick, half the height of a regular unit but with a similar depth and length, had been popular as an accent on the front façades of suburban American homes in the 1950s and 1960s. The only common properties these bricks shared with those of antiquity are their name and height. The method for manufacturing them was revealing. Double units, fired, and then split in half along a predetermined dividing line resulted in a rustic edge along the brick's narrow face. These units were the perfect element with which to distinguish the academic wing of Temple Israel in Dayton, Ohio, from the rest of the building by exploiting the shade and shadow falling on their split faces.

Previous pages: The Mary D. and F. Howard Walsh Center for Performing Arts.
Opposite: Four clay tile shapes, Temple Israel, Dayton, OH.; *above:* Frog brick, The Mary D. and F. Howard Walsh Center for Performing Arts.

Penetrating the current manufacturing culture of this centuries-old material requires passion and perseverance. Unburdening its history and connecting it to a project in a new way can result only from personal encounters with the manufacturer. Understanding the production process is important to applying the technology of manufacture in normal as well as extraordinary ways. Looking at instantly recognizable products alongside out-of-date materials that belong to an earlier period of commercial production can prove beneficial. With careful consideration, new, economical uses can often be found for these products. To determine new opportunities for materials, one must forgo the secluded, safe

office environment and venture into the world of industry and manufacturing. There may be false starts and red herrings, but there are just as many happy accidents and unexpected discoveries.

Previous pages: Roman brick and clay tile, Temple Israel.
Opposite: Brick basilica columns, Pompeii, Italy.
Left: Student Union Building, Texas Tech University, Lubbock, TX.
Following pages: Student Union Building, Texas Tech University.

08 | Appropriating materials

Something uncommon occurred in 1986, when one group of artists appropriated the work of another, and fully acknowledged using this material for their own purposes. The band, Run-D.M.C., bridged the gap between rap and rock music by co-opting parts of the Aerosmith hit of a decade earlier, *Walk This Way,* and adding it to new material. The transformed version of the song was another hit. This piece of music was a huge crossover success for Run-D.M.C. and relaunched Aerosmith's popularity. It put hip-hop at the forefront of popular music, and sampling became a major force in pop music. It is news when one artist, without acknowledgement, uses the work of another artist. Musicians have long been suing each other over songs that are merely similar, much less plagiarized, so the fact that Run-D.M.C. credited Aerosmith was refreshing and generated considerable public discussion about originality and source material.

Appropriation is common in other creative endeavors. In 1917, a work of art, *Fountain,* by Marcel Duchamp, started as a prank. He referred to the art, a urinal, as ready-made ceramic ware. Roy Lichtenstein employed Benday dots, a shading

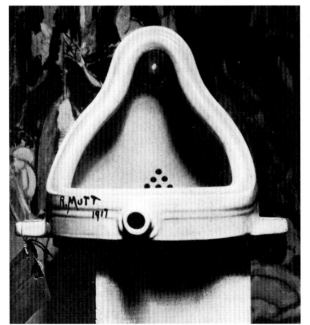

pattern normally found in newspaper comic strips, in his paintings. Andy Warhol is famous for replicating Campbell's soup cans, and the list goes on. Architects, too, often borrow, frequently from the past. Classical pediments have been in vogue for the last 2,000 years. It is difficult to imagine the stately homes of Britain, designed by the 18th-century architects, the Adams brothers, without their echoes of Andrea Palladio's Italian 16th-century villas. In recent decades debate about using old styles in new work has been prompted by the post-modern design era. Historians are quick to relate architects and styles, but architects themselves are less than forthcoming about adopting someone else's work as their own. It is rare when an originator of new material acknowledges the transference, as did Duchamp, Lichtenstein, and Run-D.M.C.

Appropriation has always been part of my work, specifically materials and systems not normally used in architecture. Products from other fields are often less expensive, and they frequently come as complete systems, having their own methods of assembly, and do not require standard contractor labor.

In 1972, the American Film Institute (AFI), in Washington, D.C., was searching for temporary quarters, an auditorium and administrative space dedicated to the presentation and preservation of film, and training in the art of filmmaking. That year, George Stevens, Jr., the director of AFI, invited me to tour the Kennedy Center to examine potential locations to fulfill the AFI's needs. For administrative spaces, we settled on an unused service corridor on an upper floor, an area envisioned as future access and support space for a theater, even though it lacked natural light. For the auditorium, we selected space originally designated as an orientation center, directly behind the Eisenhower Theater's stage house; it was acoustically isolated from the adjoining auditorium and of the same height. There were four constraints in the use of the space. First, the boundary surfaces needed to remain unaltered. Second, all AFI construction had to be removable to accommodate future uses. Third, costs could not exceed $200,000. Finally, it was imperative to schedule screenings around activities in the Eisenhower Theater since part of the AFI space served as a loading dock passageway.

The design of the film theater was straightforward. A lobby, a projection booth, auditorium seating, and a screen filled the long narrow volume. Entry to the AFI theater originated in the Hall of States, a main thoroughfare in the Kennedy Center. The lobby was compact and vertical; it brought patrons to an elevated level, adjacent to the last row of theater seats. Light steel framing, similar to scaffolding, provided a good rake for the seating and ensured excellent sightlines to the largest screen the space could accommodate. The raised audience filled half the volume of the space and focused on the screen.

During development of the design, many discussions took place about the ideal environment for film presentation with superior visual and acoustical conditions. For some projection formats, the image was almost as wide as the theater. George was concerned about acoustics; avoidance of a dead space was paramount, and electronic sound needed to be bright and lively. The two long parallel concrete-block walls, on either side of the auditorium, left in their unfinished state were an acoustical concern. Robert Hanson, the project acoustician, requested large acoustical reflectors for one of the walls, to disperse sound and to avoid echoes. To meet this requirement economically, I suggested a ready-made product from the auto industry, not the windshields from a decade earlier at the University of Toledo, but metal hoods and front quarter panels. After some deliberation and review of General Motors' detail drawings of these parts, Robert and George found the 1973 Chevrolet Impala parts an acceptable solution. The hoods and panels met all of the acoustical diffusion requirements, and General Motors donated them. The temporary AFI home in the Kennedy Center served as the showcase for this national institution for three decades. Countless film audiences visited this theater, and I am sure many of them did not know the origin of the blue acoustical reflectors.

1973 IMPALA CUSTOM COUPE

In 1970 my firm designed the Salisbury School, a modest but exciting project for this community on the eastern shore of Maryland. The cost was $250,000, similar to that of an expensive house.

The school housed classes from kindergarten to eighth grade under one simple roof. After two decades of graduating students, the institution's board decided to extend their educational program through the twelfth grade, inaugurating an upper school. I was asked to take on the assignment. The board insisted that the same unit cost, escalated to current dollars, should be the basis of the new project budget. They promised they were up to the challenge of fund-raising, and I indicated that design could commence upon program confirmation and budget verification with a local contractor.

Only some form of an economic building would conform to the monetary constraints. The client asked that the upper school's library occupy the center of the new building, just as the lower school's library did. The original building has an entry tunnel that serves as part of the daily arrival and departure ritual with the headmaster. He greets students as they arrive and sees them out at the end of the school day; the school desired a similar sequence in the new building for a corresponding ceremony.

Conceptually, a building plan started to coalesce: the library and administration at the center with four radiating arms—an entry, an academic wing, a multi-use gym/auditorium, and a vacant space for a future addition. The radiating wings, built in a conventional manner—light steel framing and flat roofs—were relatively inexpensive. The library and administration was the economic concern, as it needed to be a distinctive place, at the heart of the project.

Prior to taking on this second project at the school, I had never considered using a pre-engineered bulk storage dome, a highway structure for keeping granular material sheltered, as an architectural element. These structures, visible along roads everywhere, are hard to ignore. I soon discovered that the timber-frame structures came in a few standard sizes and basic shapes. The best part was the cost for the size required—a little more than $100,000 erected in the field by the supplier. Usually the 20-sided, 82-foot-diameter, 30-foot-high dome could contain almost 3,000 tons of salt, instead it became usable with customized finishes for the library, administrative space, and exterior skin. Wood siding, clay tile interior demising walls, special lighting, a mezzanine, and 25 octagonal windows added at minimal cost, enliven the interior. The outside, clad in standing-seam copper sheets instead of standard asphalt shingles, met the economic and aesthetic challenges while providing a vaulted space for the centralized programs.

My first large-scale project involving appropriated objects and systems occurred when, in the early 1970s, we designed a new building for the 100-year-old Brooklyn Children's Museum. The building cost slightly less than four million dollars—almost half the amount then being spent on other urban museums of comparable size and complexity—because of our firm's adroit use of resources and the museum's placement.

Five components appropriated from allied fields added to this first-of-a-kind underground museum for young people in a park setting. The exterior signage consists of typical highway placards, reflective letters on a green background, customized to announce arrival at the museum. An enameled metal Harvestraw silo encloses a fire exit as it rises above the museum's roof level into the park. A transit kiosk from the Manhattan approach to the Queensboro Bridge at Fifty-Ninth

Opposite left: 1973 Impala Custom Coupe.; *opposite right:* Impala parts acoustical reflectors, American Film Institute, Washington D.C.

Street and Second Avenue serves as the prominent main entrance for the museum. In the 1950s, train service on the bridge had been abandoned, and five kiosks, isolated on a traffic island, were still in good condition but unused. The city was more than relieved to have one terra cotta, steel, and copper-clad kiosk removed for another useful installation. For 30 years, this kiosk welcomed visitors to the Brooklyn Children's Museum. In 2005, this portal moved again, to allow the museum to expand. It now serves as an information center on Roosevelt Island, just beneath its first home on the Queensboro Bridge, but on the opposite side of the East River. Steel culverts inside the building entry descend to the four exhibition levels. This 140-foot pathway, contained within the four sections of corrugated steel culverts, defines the primary circulation through the building and guides visitors to the exhibitions. A steel storage tank encloses the largest meeting space in the building. This gathering of large-scale objects served as a complementary architectural story to the museum's collection of historic and cultural artifacts.

Shortly after completing this project, we used culverts and tanks again for another similar purpose at the New York City Fireman's Training Center (1975). The Urban Development Corporation of New York had decided to develop Roosevelt Island as a residential community, which required the existing training center be relocated to Ward's Island. The design of this nine-building training complex included a single large education building facing a series of smaller training structures. The main entries and dormers on the education building consist of tanks and intersecting steel culverts, easily located in this landscape of apparatus and training equipment.

Even though the commission for Artpark arrived in the office after the Brooklyn Children's Museum, its construction was completed several years earlier. New York State desired a public park to present the performing and visual arts and to make them accessible to everyone in a precedent-setting manner. The park surrounds a multipurpose 2,400-seat covered auditorium with an additional 1,800 lawn seats. The auditorium was under construction at the time Artpark was conceived, on a 175-acre site on the escarpment of the Niagara River, a short distance from the Falls. To meet the completion date, July 1974, the planning, design, and construction for Artpark occurred in less than two years.

The condensed schedule for construction necessitated a non-standard approach for a state-funded project. The design put forward a series of structures and enclosures clustered near and leading to the auditorium. One large element, the ArtEl, conceived as a unifying device for circulation down the steeply sloping site, provided venues for artists along its path and spectacular views to the river. The ArtEl consisted of an elevated timber boardwalk for circulation above the landscape and a plastic canopy

Opposite top: Roadside storage dome.; *Opposite bottom:* Main entry, Salisbury Upper School, Salisbury, MD.; *left:* Library, Salisbury Upper School.
Following pages: background: Abandoned subway kiosk.; *left:* Silo and highway sign, Brooklyn Children's Museum, Brooklyn, NY.; *center:* Metal culvert, Brooklyn Children's Museum.; *right:* Museum entry, Brooklyn Children's Museum.

for shelter on a series of standard roof trusses. Below this wood frame structure was an amphitheater and a town square for additional activities. A red brick pathway connects these activity areas at grade.

Since winter weather in this portion of New York State is problematic, limiting construction at the site was part of the strategy to meet the schedule. All but the two of the major architectural elements arrived at the park prefabricated and

ready for installation. Studio spaces for the artists occurred on the ArtEl in a series of small enclosures: a log cabin, an assortment of truck bodies, a metal shed, a wooden silo, and a streamlined plastic ticket kiosk. On the town square, a series of pre-engineered metal sheds and a dome accommodated similar activities. These elements provided a special festival-like atmosphere and left the majority of the acreage open for special art installations and public use. Although this portion of Artpark was built as a temporary facility for less than one million dollars, it successfully operated within and around these artifacts for more than three decades. The ArtEl was dismantled in 2005 as arts funding waned in New York State and a reorganization of the park activities occurred. A new non-profit entity now presents the performing arts events and a limited visual arts program. The state continues to oversee the park's recreational activities.

A sheet-metal product not originally produced for architectural application caught my eye as I traveled Texas highways. It is difficult to drive for an hour in that state without seeing a cattle truck. It occurred to me, after many years of encountering these bovine conveyances, that their punched aluminum panels could serve an architectural purpose.

Above: Entry culverts, Fireman's Training Center, Ward's Island, NY.; *opposite left:* ArtEl and Town Square, Artpark, Lewiston, NY.
Opposite right: Prefabricated enclosures, Artpark.
Following pages: left: Globe-News Center for the Performing Arts, Amarillo, TX.; *center:* Cattle panel soffit, Globe-News Center for the Performing Arts.
Right: Cattle panel ceiling, lobby, Globe-News Center for the Performing Arts

These panels, manufactured with small, regular openings to provide air and light to livestock during transport, now occupy points of prominence in three of my firm's recent public projects. In the Courtyard Theater in Plano, Texas, they are a wall surface in the lobby and auditorium. The openings in the panels allow for backlighting while concealing the heating and cooling distribution systems. The light washes through the pre-cut panel openings, bringing a yellow glow from the surrounding surfaces to the public spaces.

The Student Union Building at Texas Tech University in Lubbock has a newly constructed wall that encircles the public pathways to activities in this 200,000-square-foot building. Cementitious board covers both sides of this three-story wall. Two-foot-wide diagonal recesses punctuate the surface every 20 feet along its 450-foot-length. Cattle panels installed end-to-end in the recesses' full height along with a continuous band of concealed fluorescent bulbs, reflects light into the surrounding spaces. The wall unifies the interior spaces of the student union, connecting the four different eras of construction.

In the 2006 Globe-News Center for the Performing Arts in Amarillo, cattle panels are the finished underside of the billowing roof structure that encloses the lobby and outdoor terrace. First-time visitors to these buildings usually do not see the panels for what they are, because they are out of their common context. It takes a long, second look before recognition occurs. When it does, there is acknowledgement of their familiarity and their regional significance. This provocative juxtaposition of materials adds drama and a sense of informality to the surroundings.

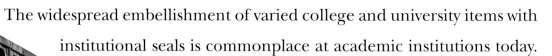

The widespread embellishment of varied college and university items with institutional seals is commonplace at academic institutions today. The circular Bowdoin College seal, a stylized sun, reminded me of a benevolent Louis XIV. Popular campus legend has it that the sun symbolized Bowdoin's location as the easternmost college in the United States (in Brunswick, Maine) when it was adopted in 1798. The seal remains unchanged after more than 200 years. In its stark but pleasing image the sun, in the center, appears to smile, while rays radiate to a Latin inscription and date surround. A little bit of yellow at the edges accentuates the strong black-and-white graphic.

I first encountered the seal on a napkin that was about to receive my drink glass, at a meeting about design of the David Saul Smith Union at the college. At the time, I was discussing the conversion of the 1912 Hyde Cage, an indoor field house for practicing outdoor events, into a 21st-century student union. In a region of the country known for its long winters, the use of natural light was an obvious need. At an early point in the design, the image of light conveyed by the college seal seemed like an obvious ingredient in the large central gathering space. Eventually a large version of the seal became a unifying floor pattern rendered in marmoleum (a type of linoleum).

Especially from the elevated levels that look down upon the lounge, the seal appears to peek out from under adjoining activity areas and the movable seating that is in constant transit to accommodate the variety of events held in the space.

From previous attempts at other institutions, I learned that using emblematic symbols could be a controversial proposition. I anticipated that appropriating the college seal for a walking

Above: Cattle transportation trailer.; *right:* Napkin, Bowdoin College seal.
Opposite: Rear-lit cattle panels, lobby, Courtyard Theater, Plano, TX.
Following pages: Marmoleum college seal, Bowdoin College, Brunswick, ME.

surface would be no different. Instead, everyone was fascinated with the use of the seal; all the building committee members wanted was a date for the completion of construction. At 90 feet in diameter, this was the largest laser-cut marmoleum installation in the world when installed in 1994. The Smith building has functioned as a "textbook" student union for the college since its opening. Its colorful, patterned, well-used interior provides a flexible environment that serves 1,000 students and visitors a day and the college seal receives increased visibility in a place central to campus life.

A half-dozen years after my Bowdoin experience, I unexpectedly discovered the University of the South seal on a napkin during a meeting on its campus. Lightning shouldn't strike twice in the same place; nonetheless, I put the napkin aside for future consideration. The development of a new dining facility was a lengthy process for this tradition-bound 150-year-old institution. Through student discussions a precedent for the formal dining area developed. The design for

the refectory, similar to those at Oxford and Cambridge, took on the appearance of a church nave, long, narrow, and tall. Composed of stone blocks quarried on the campus, an exposed wood roof structure, and glass walls looking out to the chapel, the 50-foot-high enclosure required special lighting to match the quality of the space. It was at this point in the design process that I recalled the university seal. In short order, an abstract steel and glass version of the seal became the design for three 20-foot-long chandeliers that filled the volume of the space with artificial light to brighten the evening use of the building.

My first visit to Amarillo in 1960 was quite different from my second, 40 years later. From my initial automobile trip, I remembered the windswept high plains of this West Texas city. My second visit with arrival by airplane provided another lasting impression, views of irrigation patterns. By my third and fourth excursions, I could see the Palo Duro Canyon interrupting the regular geometric agricultural patterns. The Canyon, an archaic drainage cut through the terrain, contrasted with the rigid modern circular and rectangular cultivation geometry. The green, brown, red, and purple contours of the earth flattened beyond three-dimensional recognition from 20,000 feet resembled a large, abstract quilt. The residents of Amarillo are exceptionally proud of their community and its differences from others nearby. When it was time to consider the potential use of fabrics for the 1,300 seats in the auditorium of the Globe-News Center for the Performing Arts, the irrigation patterns came to mind. This seemed to be

an ideal place for an abstract woven version of the local landscape. Four colors in the warp and two in the fill provide shades of teal and purple emphasizing the graphic configuration of the tapestry weave. With more than 600 yards of material and a very cooperative and creative supplier, Yoma Textiles, this fabric met the project budget of 35 dollars per yard.

Left: Napkin, The University of the South.; *right:* Irrigation aerial.; *opposite:* Refectory chandeliers inspired by school seal, McClurg Dining Hall, University of the South, Sewanee, TN.

Above: Copper reflectors and metal grating ceiling, Performing Arts Center, Texas A&M University-Corpus Christi, TX.
Opposite: Irrigation pattern fabric, Globe-News Center for the Perfoming Arts.

A full and proper introduction to the island of Oahu, Hawaii, requires a 15-minute detour from Honolulu nightlife in Chinatown or Waikiki to the edge of the Punahou School, on a nearby hillside. A low wall, fully draped in night-blooming cereus, separates this gently sloping century-old campus from its neighbors. My first visit to the island coincided with the once-a-year blooming season of this plant. On that night, in 1990, its white flowers were bright against its dark, shiny, green stems and the supporting stone wall. I was in Honolulu to inspect its last remaining movie palace, the Hawaii Theatre, in the Chinatown section of downtown and to assist in determining the feasibility of its re-use as a performing-arts center. My hosts were doing their best to show me this tourist paradise, including its flora and fauna. Seeing this unusual plant didn't register as strongly as visiting Bertram Goodhue buildings or Pearl Harbor, but I later had occasion to remember this brief evening excursion to the school.

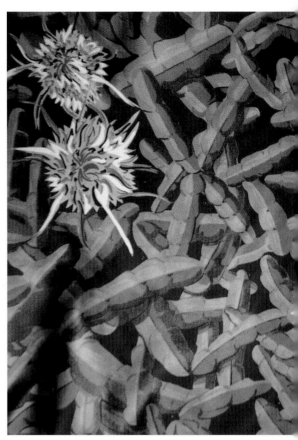

My second commission in Honolulu started a year later, at the Punahou School itself. Updating Goodhue's 1926 Dillingham Hall was a challenge. He was well known for designing the Nebraska State Capitol, the Chapel at West Point, and other religious and academic structures. For more than half a century Dillingham served as a campus assembly space. During World War II the United States Army Corps of Engineers occupied parts of the Punahou campus. Dillingham Hall served them as a drafting department by extending a flat floor over the orchestra-level seating and building a second floor extending from the balcony to the stage. By the time I arrived, these changes were long gone; though returned to its previous form, it barely functioned as a performance facility. A long list—improvements to the infrastructure, a side-stage addition, toilets in a separate structure, and careful restoration of original details—were all part of this intricate project.

Left: Sketch of night-blooming cereus.; *below:* Wall of night-blooming cereus, Punahou School, Honolulu, HI.

Upon Goodhue's unexpected death in 1924, Hardie Phillip, an associate in his firm, completed the design and construction of Dillingham Hall. While working on the restoration of and addition to Goodhue's Los Angeles Public Library a few years earlier, I had gained a first-hand appreciation of his decorative approach to design. His usual stencils, wall reliefs, decorative art items, and colorful surfaces were missing from Dillingham's interior, which appeared nude in contrast to his mainland projects. I assumed that his early death prevented full development of the design. Because Dillingham is an historic structure, every addition to the building was conceived as a removable element for future generations. The items I introduced are distinct insertions in the interior and discrete exterior additions. My intentions were to produce new functional items in the auditorium and to complete the original design in the spirit of Goodhue's other work.

My initial tour of the island came to mind when it was time to replace the stage curtain. The night-blooming cereus was a symbol of the state but also of the school. Many theatrical and musical events in the revitalized hall would take place in front of the stage curtain as well as behind it; to make a visual feature of this surface would heighten the experience of attending a presentation. A small magic-marker sketch of a night-blooming cereus on a dark blue background evolved into a full-size drawing and eventually a velour fabric sample. I was delighted that the school building committee immediately recognized the drawing image and approved its use for the stage curtain. As Goodhue might appreciate, the new theatrical- and house-lighting bridges, the vertical air-distribution system, and the proscenium wall all take their color palette from the stage curtain. It is presumptuous of a current architect to assume he is completing the design of another architect by appropriating his style, but I enjoyed this unauthorized collaboration with Bertram Goodhue and Hardie Phillip.

Observing more than just the buildings around us, architects can find inspiration in the most unexpected or mundane objects and materials found in all spheres—from industry to landscape to transportation to art—and should feel no constraint in using them as their ability to function equally well in the built environment becomes apparent.

Opposite: Stage curtain, Dillingham Hall, Punahou School.

09 | Sustainability

In 1991, when a friend telephoned to tell me about his spouse's respiratory illness, all I could do was listen. I learned that the old prairie-style house they lived in had contributed to it. Though indiscriminate modernizations performed during the preceding decades prevented determining its exact source, it appeared that mold, in combination with a forced-air heating and cooling system, had been a major factor. The residence was aesthetically appealing, but they had decided to take up new quarters and asked if I could design a healthy house for them. I said yes, and thus began creation of a specialized residential environment.

From this urgent inquiry, the Highland House, in Madison, Wisconsin, evolved the next year. The design process began by defining what a healthy environment was, exactly. Many of the ground rules are commonplace today, but in the early nineties, one had to conduct considerable research. The clients' fixed goals became design necessities—natural materials, orientation for maximum daylight and minimum heat gain, shading for glazed areas, water-based paints, a building frame constructed from solid wood members, natural ventilation, filtrations systems, and sun-dried floor tiles were among the basic elements. Based on conditions at the existing residence, a list of don'ts also developed—no carpets, to minimize dust and mold collection, no glue as an adhesive, and no synthetic materials or finishes, to prevent off-gassing. A prominent landscape feature, a specimen American chestnut tree not usually found this far north in Wisconsin and a survivor of the blight, was a major consideration in arranging the new house on a site in a well-established neighborhood.

Since this project, environmental considerations and sustainability issues have been integral to my architectural practice. Client goals for sustainability are established at the outset of every new undertaking. The intelligent use of materials and the proper application of resources is an essential component of architecture. The impact of construction on the environment has led architects to develop a series of sustainable standards, among them the use of local and regional products, recycled and low-embodied energy materials, and resource conservation. My architectural practice has recycled entire buildings by adapting them to new uses and has led the profession in this arena. In 1980, we received one of the early awards from the American Institute of Architects for transforming a moribund department store and movie theater

Opposite: Madison Civic Center, Madison, WI (demolished 2005).
Following pages: Highland House, Madison, WI.

into the Madison Civic Center, in Madison, Wisconsin. Our design defined an innovative way to allow humble but useful existing structures to serve new functions.

The standards developed by architects for sustainable and healthy environments are codified by members of the U.S. Green Building Council (USGBC) into LEED (Leadership in Energy and Environmental Design). These criteria apply to new construction and renovation projects, existing building operations, homes, and neighborhood development. It was created to establish a common standard of measurement for a "green building," promote integrated, whole building practices, and transform the construction market.

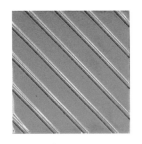  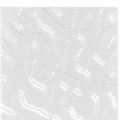  LEED provides a framework for assessing building performance and meeting sustainability goals, it emphasizes strategies for sustainable site development, water savings, energy efficiency, materials selection, and indoor environmental quality. It recognizes and promotes expertise in green buildings through a system offering project certification, professional accreditation, training, and practical resources. Some clients choose not to use these standards but instead follow an independent course and establish their own standards. The '2030 Challenge' is a new global architecture initiative targeting development projects to use half the fossil fuel energy a typical undertaking would consume. The American Institute of Architects has endorsed and adopted this enterprise for resource conservation, sustainable design, use of water resources, and the certification of products, among other things. The relationship of materials to these programs is instrumental to their success.

ImaginOn is the first building in Mecklenburg County to receive LEED Silver certification from the USGBC and one of the first in the state of North Carolina. Designing this new facility in Charlotte was an excellent opportunity to build an exemplary community project. ImaginOn brings together the existing programs of the Public Library of Charlotte and Mecklenburg County and the Charlotte Children's Theatre. By revealing itself as a series of distinct elements, the architecture serves as an educational tool and provides a new chapter in the written, spoken, and electronic story-based program of activities. The building design optimizes energy performance while reducing energy consumption. Daylighting, provided by perimeter glazing and north-facing clerestory windows above the roof, enhances interior spatial qualities and emphasizes the main public circulation route. At the same time, it reduces dependence on artificial lighting. Generous roof overhangs and window-shading also modulate direct light while minimizing unwanted heat gain. Throughout the building, specially selected, rapidly renewable and recycled content materials, recycled clay, recycled stone, regional, and highly durable materials are used. These are low-emitting materials for improved indoor air quality.

One material in this project is out of the ordinary. Elsewhere (Chapter 3), I have noted a long-standing, working relationship with the tile manufacturer Tom Sawyer. For this project, he produced 40,000 square feet of 6-by-6-inch tiles that are different from those employed on my previous projects. By using a standard tile size and only changing the surface, there was a minimal cost for new dies. One of three new surface patterns—linear, diagonal, and vermiculate rustication—runs across the face of each tile and completes itself on the next; this results in a change in the scale of the completed wall surfaces. While discussing the fabrication of new dies to make the tiles, we decided to use greater quantities of recycled material. This first-time undertaking was edifying for everyone involved. Normally some recycled material is used in making most new clay products. For these tiles, 32 percent of the material was post-consumer recycled glass and 16 percent was post-industrial grinding paste from the computer industry. We found that controlling shrinkage while firing the recycled material with the clay was harder to manage than

anticipated. In turn, it was difficult for the installers to align the tiles on the walls with the desired tolerances since we specified minimal grout joints. In fact, the tile had to be installed a number of times until it was satisfactory. The

Opposite: Samples, three types of high-content recycled clay tiles.; *above:* Tile enclosures, ImaginOn: The Joe and Joan Martin Center, Charlotte, NC.

completed tile wall surfaces penetrate the exterior glass skin of the building. The three distinctly colored and patterned surfaces are one of the characteristics of ImaginOn easily identified from a distance. Now that we have accomplished this first project, we can envision future applications for this material and ways to control the tile shrinkage and improve the field installation.

The George A. Purefoy Municipal Center in Frisco, Texas, a rapidly growing area north of Dallas, is designed to the LEED Silver level. The prototype for this civic structure is the traditional Texas town green and courthouse found in most counties across the state. These older structures are masonry, and the Frisco community wanted the new City Hall to be constructed from Texas stone, with the stipulation that taxpayers not be overburdened. Local products are desirable for rooting a building in its community and minimizing travel distance and shipping costs.

Above: George A. Purefoy Municipal Center, Frisco, TX.; *opposite:* Granite column covers, George A. Purefoy Municipal Center.

Numerous types of limestone are readily available in Texas, though its color varies by location. Judicious selection based on the size of stone blocks can make it an affordable and acceptable product. Granite is another matter; there is only one source for this stone in Texas, and it is expensive to process and finish. The City Hall's front façade incorporates two colonnades on either side of a central entry tower facing the new town green. Granite was the material of choice for both the colonnade and the entry tower to distinguish them from the limestone and glass cladding of the remainder of the building. This composition focuses attention on the main entrance from Frisco Square. At the outset of the consideration of materials, I knew it was possible to afford limestone, but granite was questionable.

Stone-cutting equipment developed in 1980s has led the way to a major resurgence in worldwide granite production. Fabricators utilize this equipment to make stone sheets just ⅝-inch thick in some cases, for two very different applications—lightweight

exterior façades and residential countertops. The fabrication process for this thin sheet material created a new form of waste. Oversized granite sheets normally produce a 2- to 4-inch remnant strip from each side of the granite sheet. This scrap material now joins other remnants in the spoil piles in fabricators' yards.

During the project's schematic design phase, we explored using two types of affordable granite blocks for the City Hall's central tower: 8-inch-high split-face units and 30-inch-high by 8-inch-thick pieces, fabricated from the skins of quarry blocks. Once the client reviewed these two options in a mock-up of the exterior walls, they decided on the split-face units.

The 28-foot-tall by 5-foot-diameter columns were unaffordable in small granite blocks, or blocks of any size, so we turned to the option of using the thin sheet remnant material. Discussions with a precast concrete manufacturer and the granite supplier resulted in development of a potential fabrication method. By aligning 4-inch scrap material end-to-end, narrow face down in a form and then pouring a concrete backing to adhere the pieces together, we were able to make self-

supporting units up to 14 feet in height. Casting one-quarter of a column circumference at a time allowed it to join identical pieces to complete the column shape. The granite, formed with spaces running between the pieces, visually recalls traditional column flutes. A 6-foot-tall mock up, approved at the precast plant, led to an affordable granite column cover. A later mock-up at the construction site allowed the owner to approve the column and our team to establish a standard of quality for the fabricators and installers.

The processing of stone varies from quarry to quarry. Since inauguration of construction on its Lubbock campus in 1923, Texas Tech University has used stone only from Lueders, Texas. My first visit to these quarries was in 1999, while

Above, left to right: Granite countertop fabrication; Granite scrap; Granite scrap in concrete form; First mock-up of precast granite column.
Opposite: Turkey track, Lueders Limestone.; *inset:* Turkey track bungalow.

designing the fourth addition to the campus Student Union Building. The expanded building contains a bookstore, student government and organizations' offices, lounge space, food service, and a gathering pavilion, a unique multi-story campus space for informal events. Our objective was to find stone to match the original portions of the building. We not only met this objective but also found a material that added to the project's success.

There are many small towns between the Lubbock campus and the quarry. Grain silos, a few large houses, bungalows, shotgun residences, and commercial strips unfolded along the drive. However, in the community of Lueders we saw a number of arresting residential buildings made of stone. These small homes, built from remnant material, were typical of many quarry towns across America with one striking exception. The stone surfaces on these residential buildings "wiggled." We stopped the car to examine the façades of a few structures. Recording this phenomenon used up an inordinate amount of digital memory.

The representative from the stone fabricator met us in town and told us that the material on these buildings was "turkey track." We looked at the stone again—the tracks that turkeys make? For us, this comparison was a visual stretch. We drove out to the quarry to find matching stone and the source of this unusually named material. Lueders stone lies in 14-to-16-inch ledges, quarried in 4-to-8-foot lengths until they are exhausted in a given location. Then the operation moves to another portion of the ledge to start again. Near each excavation area, a scrap pile of turkey track was evident. The discarded stone was the typical uniform beige material but one side contained fossils; the turkey track pattern came from the very bottom of the ledge. Two different creatures formed this pattern 360 million years ago, when the area was a seabed. We subsequently learned from a university geologist that the "tracks," produced by worms and a shrimp-like crustacean, had filled with sediment as the limestone deposit formed. Due to the irregularity of the fossils this portion of the limestone was not used except by the local residents of Leuters in search of an inexpensive construction product.

A sample of this material returned with us to the office for design consideration. The student union project had a fixed budget that allowed for the use of some simple stone blocks. We could not afford the decorative hand-cut shapes employed on the original 1920s campus structures. We presented the turkey track as an ornamental accent to the flat-cut stone surfaces to the building committee as part of the schematic design; no one present had seen it before. With unanimous approval by the students and building committee, the turkey track became a series of decorative bands on the piers that form the gathering pavilion. The most important and popular corner of the enlarged student union is now the official campus home of some of the oldest Texas fossils permanently on display, except for samples in the geology building. This material might have ended up in a spoil pile if we had not visited the stone quarry.

Our second project on the northern New Jersey campus of Ramapo College, the Spiritual Center, is unlike other recently constructed buildings, even the Berrie Center for Performing and Visual Arts that we completed on campus in 1999. The Spiritual Center is a privately funded project at a public state college. It will provide a centralized location for

students and faculty in times of crisis as well as during periods of celebration. This sanctuary from daily academic life will be available to organized religious groups and individuals for spiritual use 24 hours a day. A series of small structures, totaling 2,000 square feet, sits at the southwest edge of the college pond, a unique campus landscape feature. The sanctuary, the largest single space for as many as 80 individuals to gather, takes advantage of solar orientation with a large north-facing clerestory rooftop opening in addition to windows facing the pond. Through their arrangement, two small, 6-by-10-foot, quiet meditation spaces, the sanctuary, and a service structure define an outdoor gathering space. Natural ventilation, zoning of each structure for independent use, daylighting, and a high-performance heating, ventilating, and cooling system provide some of the means to minimize and balance energy consumption. When complete it will serve as a model for other campus projects to follow.

The intelligent and economic use of materials and resources is an essential component of architecture. Building with sustainable, recycled remnants and renewable materials permeates my firm's current projects, even those with limited budgets. Although product selection is only one facet of our design process, it has vast implications for a building's end result, and it is essential that architects stay abreast of products available both near their projects and nationally in a quickly evolving marketplace.

Opposite: Gathering pavilions, Student Union Building, Texas Tech University, Lubbock, TX.; *above:* Rendering, Salameno Spiritual Center, Ramapo College, Mahwah, NJ.

10 | Art and architecture

Visiting Fifty-seventh Street carpet showrooms in search of flooring material for the new Best Products Corporate Headquarters proved unrewarding. In 1978, after a year of design work, the project was about to start construction and it was time to complete the selection of the interior finishes. Frances Lewis, a founder of the company, and I were looking for a patterned carpet to use in the open office areas and secondary circulation pathways through the building.

In a considerate way, Frances let me know that what we had seen that afternoon was unsatisfactory. I should have known this would be the case, because Best Products and the Lewis family had a significant collection of contemporary art, an international assemblage of 19th-century decorative-art pieces, and 40 mid-20th-century jukeboxes. A modern rendition of a 200-year-old floral carpet pattern was of little interest to Frances, and upon reflection, I, too, found this imagery inappropriate to the project.

Floral carpets did, however, grace some of my other building projects. Their large patterns absorbed abuse and camouflaged wear, while exaggerated colors animated the spaces they occupied. The standard fare for office carpeting at the time was solid gray or beige, no pattern. These looked good for a few months following installation, but not much longer, and I shied away from selecting them.

Following our last showroom stop, it occurred to me that creating a new floral patterned carpet would be not only possible but also desirable. Frances was immediately enthusiastic. Both of us were familiar with a recently completed Japanese-style, four-part folding screen exhibited in museums and galleries across the country by Jack Beal, the realist painter known for his striking compositions. I discovered that the Lewis family owned one of these screens, depicting a lily pond with a rowboat, in colors ranging from gray to orange. By the end of the afternoon, we had agreed that I would pursue development of a custom carpet using the Beal folding screen image as a starting point for the pattern.

Above: Jack Beal, *Rowboat*, 1977, silkscreen on four panels, 71½" x 95½".; *opposite:* Samples, hand woven carpet based on Jack Beal's *Rowboat.*

After a close examination of the screen in Beal's Tribeca loft, I thought one of the end panels, with just lily pads, flowers and water, would be a good place to begin the carpet design process. Jack and his artist wife, Sondra Freckelton, liked the idea and volunteered to work with me in refining the pattern and colors. I left their loft with an extra print of the screen's end panel rolled up under my arm.

Later, in my office, after closer examination I determined that this portion of the original screen could serve as a basis for a new, enlarged carpet pattern. A repeat of just the water lilies, the width of the carpet roll, could be generated by mirroring the original image, flipping it from left to right, and mirroring it once again, flipping it from top to bottom. This became the basis for a 12-foot-wide carpet roll with a 4-foot repeat in the pattern. The dimension of this newly generated image corresponded to the scale of the headquarters building.

Discussions followed with Jack and Sondra to review this proposal and then to slightly modify the colors used in the original screen. The finished carpet would receive heavy traffic and the colors would dull over time. Intensifying the blue water, green lily pads, and orange highlights provided more contrast and compensated for anticipated wear. With the Lewis's blessing and an acceptable pattern, I approached the Mohawk Carpet Company, based in Amsterdam, New York, about producing it. During the previous decade, I had established a good working relationship with its New York City representative. At our first meeting to discuss this new undertaking, he assured me that the quantity of material required would allow Mohawk to manufacture the carpet at the cost of readily available standard ones.

Before manufacture began, I became familiar with the method of proceeding from original art to final roll goods. Mohawk staff began by placing a small 12-inch-square grid over the artwork, creating vertical and horizontal rows for each inch of the carpet (nine rows to the inch is a tight grid). Each space in every row of the grid received a number, a specific color corres-

 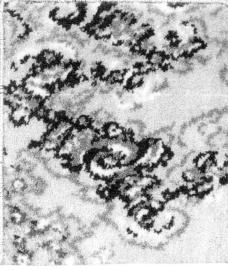

ponded to each number, and a paper "strike-off," a color drawing, resulted. A detailed review of the strike-off by Jack and Sondra confirmed the accuracy in translating the image to the gridded colors. A few were intensified and the strike-off returned to Mohawk. From this, they prepared a 12-inch-square hand-woven yarn sample. Until the manufacturing process started, all pre-production samples were handmade to refine and review the design.

2/5   "DALLAS 14 JACK 6"   Red Grooms

Finally, after months of development and review, it was time to make the carpet. To remain on schedule for the opening of the building, a final approval meeting was required. As often happens Jack and I agreed to meet at our first free moment, a Sunday afternoon in late January. I didn't realize until arriving at the Beal's loft that this was the day of an annual American ritual, the Super Bowl. The game had just started, and a group of their friends were gathered around the TV to witness the spectacle. While Jack and Sondra reviewed the final hand-woven color sample, the group of enthusiasts cheered the teams to greater heights. Like everyone else, Red Grooms, another distinguished artist, was watching the game, eating, drinking, and observing as the carpet colors improved. At the same time, Grooms was preparing an etching plate of Jack watching the football game. The finished print would later be titled *Dallas 14 Jack 6*. By the game's end, the work on the image for the carpet was finished, as was Red's etching. It took a lot less time to complete the etching than the carpet design!

Up to this point in the process, the samples prepared by Mohawk had been hand made, each requiring about a month to generate. The costs to set the looms up for a small sample were prohibitive until production was ready to start. The first machine-made sample, reviewed a few weeks later, looked excellent. Following approval by all parties involved, the roll goods went into full production. Until 20 years ago, the design work and samples for a new carpet pattern were hand generated, and it took a considerable amount of time, in this case four months, to prepare and review samples. Today, thanks to computers, initiating design and authorizing changes can be done easily and quickly in about half the time. While the back-and-forth process remains much the same, it now occurs electronically and often without as much need for personal contact between the individuals involved.

The Best Products project was my first encounter with making an original carpet. I have gone on to generate many others. Subsequent inspirations for their patterns have come from varied sources: art, other architects' designs, decorative images, and landscapes. For the addition to the Los Angeles Public Library, my firm developed a carpet inspired by the early 20th-century architect, Bertram Goodhue, known for his integration of decorative arts into his buildings. Goodhue's stencil patterns for the ceilings of the original 1928 structure inspired our carpet designs for the 1993 addition. For the Evangeline Atwood Concert Hall in the Alaska Center for the Performing Arts in Anchorage, an abstract wall hanging by the artist Jon Friedman served as the inspiration for the aisle carpeting.

This installation was one of more than 25 projects designed by artists, which ranged from marquee lighting to lobby windows.

Best Products, the first time I worked directly with artists in the production of an art piece for one of my buildings,

was a collaborative process streamlined by the clients and my earlier acquaintance with the artists. From this remarkable undertaking, I concluded that incorporating art into the fabric of a building could intensify a memorable place. I have worked with other artists on a variety of projects since. All of the results have been gratifying, but no others have allowed a client to "walk on water" as this one did at the Best Products Corporate Headquarters.

During the design process for the Best headquarters, I became accustomed to Sydney Lewis's questions. He often used direction by inquiry to move a discussion forward. One day, halfway through the design process, I received a phone call from Sydney, who asked if I was familiar with the stone eagle sculpture that had adorned the former East Side Airlines Terminal

Opposite: © 2006 Red Grooms / Artists Rights Society (ARS), New York, Red Grooms, *Dallas 14 Jack 6.*; *left:* Reading room, Bertram Goodhue stenciled ceiling and new carpet, Los Angeles Public Library: Central Library, Los Angeles, CA.; *above:* Beal-inspired carpet, Best Headquarters, Richmond, VA.

building, on Forty-second Street and Park Avenue, and if I would be interested in incorporating it into the design. Demolition of the Terminal building occurred in 1976 to prepare the site for a new Philip Morris headquarters. I knew the sculpture, as did most New Yorkers, but I thought it had been destroyed with the building. The eagles had crowned this small structure, diagonally across the street from the sculpture of Mercury, atop

Grand Central Station. By comparison to Mercury, a prominent element in a very large composition, the eagles had been the most interesting and distinctive part of this 1940-streamlined structure. My immediate response to both of Sydney's questions was yes, even though I did not know the eagles' whereabouts.

Sydney went on to inform me that the sculpture was now in a warehouse in Carlstadt, New Jersey. When the building was demolished, the eagles had been removed from their rooftop position, crated, and placed in storage until another potential location could be found. A Philip Morris representative had informed Sydney that the sculpture was available. Since design for the headquarters building was not complete, I indicated that the eagles would be welcome in the project but that I would need time to determine how to incorporate them.

A quick trip to the warehouse was organized. Close inspection of the sculpture revealed that care taken in removing it from the building had left the 40-year-old limestone sculpture in good condition; weathering in the New York City atmosphere had little overall effect on the massive stone blocks. There didn't seem to be any physical reason not to use it.

A month after our conversation, I met with the Lewises and the

Opposite left: Goodhue-inspired carpet runner, Los Angeles Public Library: Central Library.
Opposite above right: Goodhue-inspired carpet sketch.
Far left: Carpet, Atwood Concert Hall, Alaska Center for the Performing Arts, Anchorage, AK.
Left: Wall hanging, Jon Friedman.

building committee to discuss design topics and the potential use of the eagles. In the interim, I had researched the work of the eagles' creator, the artist René Chambellan, whose distinguished career had been devoted to architectural sculpture. He worked with the famous early modern architect Raymond Hood on the Chicago Tribune Tower, the New York Daily News building, and on various projects at Rockefeller Center, including the Channel Garden Fountain heads and decorative bronzes. He did other prominent New York City projects with Sloan and Robertson at the Chanin Building, and with Cass Gilbert on the New York Life Insurance building. The eagle sculpture consisted of two stylized birds arranged tail feather to tail feather, joined by a common wall and surmounted by a flagpole and two lantern-like light fixtures mounted on their backs. The entire composition, originally viewed by travelers from 50 feet below as they exited Grand Central Station, was intended to be seen as a silhouette; the profile of the heads, beaks, and stylized wings all suggesting flight. Even so, Chambellan sculpted the birds as fully realized three-dimensional objects.

At the end of the design review meeting, I used a small cardboard model to indicate the location for the sculpture. I proposed separating the eagles at the tail feathers (removing the wall originally joining them), and placing them 25 feet apart on the ground to form a portal 30 feet from the headquarters building's entrance. Everyone entering and leaving the headquarters would get an up-close view of the sculpture while traversing this open portal. When lit during the evening the restored lamps on their backs would serve as a beacon to those on the approach road and in the parking area. Although the alteration to the original piece was dramatic, it met with the committee's acceptance. Negotiations followed to secure the eagles, and in short order they were cleaned, restored, and installed, to be seen by a new generation of admirers in a new location.

This sequence of events proved to be the opposite of working with Beal and Freckelton on the design of the carpet. In this instance, the artist, dead for two decades, was not part of the process. However, considerable care went into siting the sculpture to enhance its

Left: Rene Chambellan eagle sculptures, Airlines Building, New York, NY.

appearance and to make it an integral part of an architectural setting. With the success of salvaging this great New York City artifact, I knew exactly how to react when I learned that the original elevator cabs from the RCA (now GE) building at Rockefeller Center awaited demolition. Sydney was as positive in his response to my question about the elevator cab

as I had been to his about the eagles. I managed to secure one mahogany cab, which, restored and installed, transports employees and visitors to the headquarters' two floors. There was initial confusion in using the elevator in its new location due to the cab's original placement within a specific elevator bank at the RCA building. The elevator indicators designate floors 41 to 55; current passengers have to select 42 to get to their destination on the second floor, a small inconvenience for those who use this elegant piece of mid-20th-century cabinetry.

Indirectly, the West Wing of the Virginia Museum of Fine Arts provided another opportunity to continue the integration of art into the fabric of a building for the Lewises. In 1992, Sydney and Francis Lewis and Mr. and Mrs. Paul Mellon donated art collections and funds to the Commonwealth of Virginia for the construction of this addition. As the design developed, a need arose for two portals to the building. I suggested commissioning the sculptor Albert Paley, known for forged and fabricated metalwork, to design the gates for these openings to the building. I met Albert a decade earlier, during a lengthy wait for admission to a design review meeting of the Pennsylvania Avenue

Above: Best Products Company Inc. Headquarters.; *following pages:* Service gate, West Wing, Virginia Museum of Fine Arts, Richmond, VA.

Development Corporation, in Washington, D.C., while presenting revisions to the competition design proposal for the Willard Hotel. There I learned of his collapsible tree grate proposal, which allowed for tree growth, a significant breakthrough for this ordinary piece of street hardware. Its breakaway steel rings provide for expansion of trunks and

roots over time. In succeeding years, I watched these grates move in slow motion, as the trees along Pennsylvania Avenue matured. After this initial meeting I visited a number of Paley's larger commissions and his Rochester, New York, studio.

The massive rolling steel gate he created to screen the Virginia Museum's service yard from the adjacent sidewalk and neighborhood is like no other. Its scale responds to the large, rusticated, limestone blocks that form the base of the West Wing and the service yard. He created a series of steel staves, 11.5 feet tall and 12 inches wide, welded at a diagonal along the 32-foot length of the gate. Its appearance can be likened only to a contemporary version of a stockade fence. Originally conceived as individually forged steel staves, they were finally fabricated from formed sheet steel to reduce the overall weight of the gate. The staves taper at the top, allowing light to filter through and animate the metal of the gate, echoing the way light falls across the adjacent stone façade. This 7,500-pound gate can move at the touch of a button, belying its sturdy appearance.

When I discovered that the last remaining section of New York City's West Side Highway was on the verge of demolition and that some light stanchions were available, I immediately phoned Sydney Lewis. Our conversation was similar to the one about the elevator cabs; I was again the one asking questions. His answer now graces the west side of the Virginia

Museum. Two retrofitted light stanchions symmetrically frame Paley's other gates on the West Wing's park side, its only access point along this façade. In hindsight, I now recognize the ease of these transactions. Today I am more likely to be consulting with an art-selection committee than collaborating with a single, fast-acting, decisive donor.

Albert and I have integrated his work into subsequent projects. The continuing success of our partnership resides in his understanding of the architectural conditions surrounding his contribution

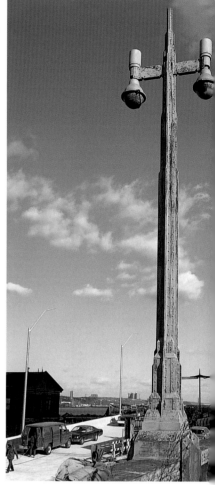

and the development of artistic responses. For example, the focus of the sanctuary at Temple Israel, in Dayton, Ohio, is on the bima, the raised altar area. Every seat in the sanctuary looks toward the rear of the platform and the ark, which contains the Torah. A series of receding wood planes constructed of poplar boards terminates in an 8.5-by-7.5-foot opening. Albert filled this opening with a patinated steel parting screen that depicts an abstract version of the burning bush. It adds to the recession of surfaces on the bima and to the perspective of the larger room. The screen, the space's focal point, provides a backdrop for all events on the bima. Once the screen and eternal light were in place, the congregation was so pleased that they commissioned two chandeliers by Paley to complement the initial installation.

In 1998, Texas Tech University, in Lubbock, commissioned my firm to expand and improve the student union. The university had a public-art manager, not a selection committee or donor. Her appointment was made as the project started construction, in the summer of 2001. By the time we met for the first time, the selection of artists was underway.

Above: Albert Paley welding sections of the service gate, West Wing, Virginia Museum of Fine Arts.; *center:* Light stanchion, West Side Highway, New York, NY.; *opposite:* Re-installed light stanchions, West Wing, Museum of Fine Arts.

She had commissioned Terry Allen, a one-time Lubbock resident who is well known for his music, songwriting, and sculpture. We discussed a list of candidates she wanted for the second commission, which included Tom Otterness, the New York-based, Wichita-born artist known for his whimsical bronze sculptures around New York City and the rest of the country. I volunteered to contact Tom, because we had developed a project together for the Eastman Reading Garden at the Cleveland Public Library a few years earlier.

Tom was selected by the university, and shortly thereafter a meeting was held in my office to formulate an approach to both Terry's and Tom's projects. The artists were familiar with each other's work, but had not met. A long, circuitous discussion began, and by the end of the day, everyone agreed that the two projects should relate. Terry was interested

Opposite: Ark Screen, Eternal Light and *Wall Sconce*, Albert Paley, Temple Israel, Dayton, OH.
Above left: Albert Paley welding *Ark Screen* for Temple Israel.; *above right:* Tom Otterness and Malcolm Holzman, Eastman Reading Garden gate, Cleveland Public Library, Cleveland, OH.

reading
book

carrying
books

FIGURE
MADE
OUT OF
BOOKS

standing
on books

in creating a *Bookman,* a figurative work composed of cast-bronze books. Tom described several possibilities, but the one proposal everyone liked was *Tornado.* His piece would sweep up a few of the *Bookman's* books, some artifacts from the adjacent library, and detritus from the student union. It would provide a link between the sculptures and the buildings while commenting on the extreme local weather conditions.

Some weeks later, both artists presented a lecture to students and faculty on campus and then visited the construction site. Otterness decided a good location for his piece would be directly in front of a new entrance to the student union and Terry decided the *Bookman* could be halfway between *Tornado* and the library entrance. Tom also came away from this visit with a second commission from Tom Shubert, the director of the student union. The school mascot, the *Masked Rider,* was to be located somewhere in the building.

After months of contractual negotiations—art and contracts have difficulty mixing— the art pieces began to take form. Terry quickly developed a maquette for *Bookman.* Tom's start was slower. I visited his studio in New York City a number of times as he considered *Tornado.* The cardboard mockups for this sculpture resembled a dust devil—a small tornado. I was concerned that it wouldn't be large enough to hold its own in the wide-open space between the buildings, even with new landscaping in place. Eventually, *Tornado* grew in height, though the budget remained unchanged.

Terry confirmed the site for his piece the following summer, 2003, and his project moved quickly forward to completion. Tom visited the site again during one of my regular visits to observe the progress of the building's construction. Cardboard mock-ups of his two pieces, *Tornado* and *Masked Rider,* were evaluated in various locations. *Tornado* was sited on axis with a new entrance to the building, as originally contemplated. Making it bigger

Above: Bookman sketch, Terry Allen.
Right: Tornado sketch, Tom Otterness.
Opposite: Bookman and *Tornado* sculptures, Student Union Building, Texas Tech University, Lubbock, TX.

TORNADO OF IDEAS

had been a good decision, even though Tom's budget was getting tight. The ideal location for *Masked Rider* was less obvious, and discussions brought to mind the most impressive equestrian sculpture I had seen in recent years. Carlo Scarpa, one of Italy's most prominent architects in the second half of the 20th century, had transformed the Castelvecchio, a Renaissance fortification in Verona, into a museum. Scarpa removed, added, and restored parts of this structure, which dated from the 14th century, and also oversaw the installation of the art in the completed museum. The equestrian statue of Cangrande Della Scala commemorates one of the city's distinguished leaders. Scarpa considered a number of locations before settling on the diagonal orientation of a cantilevered platform approximately 30 feet in the air at one end of the building, allowing observation of the piece from many vantages, all of equal importance. As a tribute to Scarpa, I suggested placing *Masked Rider* at the edge of a balcony cantilevered over the largest public space in the student union. Upon my return to the office, I sent an image of

the Scarpa installation to Tom, and he agreed to consider this placement.

When it was time to install his piece, Terry decided to move *Bookman* for greater visibility, away from the library and *Tornado,* closer to a roadway bus stop adjacent to the student union. A wry, subversive irreverence marks Tom's finest installations. *Masked Rider,* especially, delights me because it undermines not only its own importance but also that of most equestrian sculptures. Tiny figures in prominent locations hold up one end of *Masked Rider*'s stone and bronze base. Visually, the piece challenges gravity and suggests that it might tip off its balcony roost onto the floor below. Miniature Masked Riders, commissioned by Tom Shubert, have become bestsellers at $27.95 for alumni and $17.95 for students, in the union's bookstore.

Artwork designed for a specific context loses the fragility often associated with gallery and museum installations, appearing instead securely anchored in a deliberately designated space. As these projects indicate, joining art and architecture can result in something more than independent objects placed next to each other; the two disciplines can fuse, so that we look at the art, the materials, the architecture, and all that surrounds them in new ways.

Opposite: Masked Rider, Tom Otterness, Student Union Building, Texas Tech University.
Above: StoryJar, ImaginOn: The Joe and Joan Martin Center, Charlotte, NC.

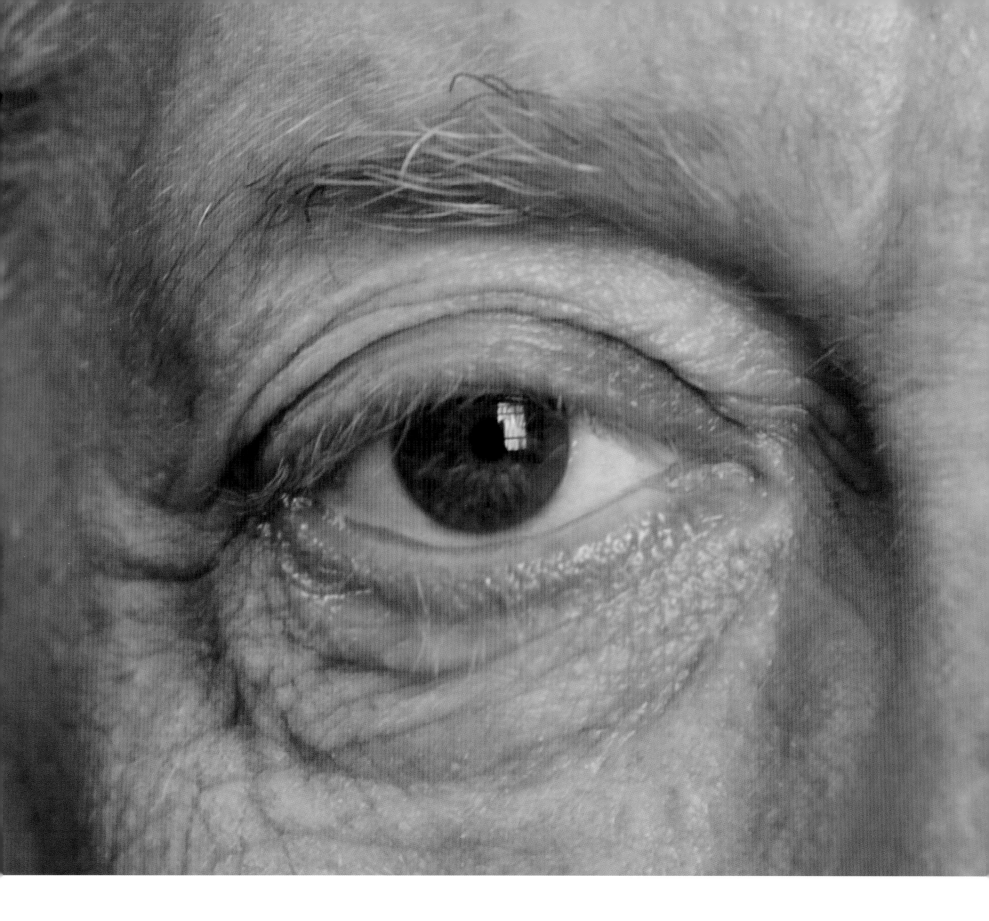

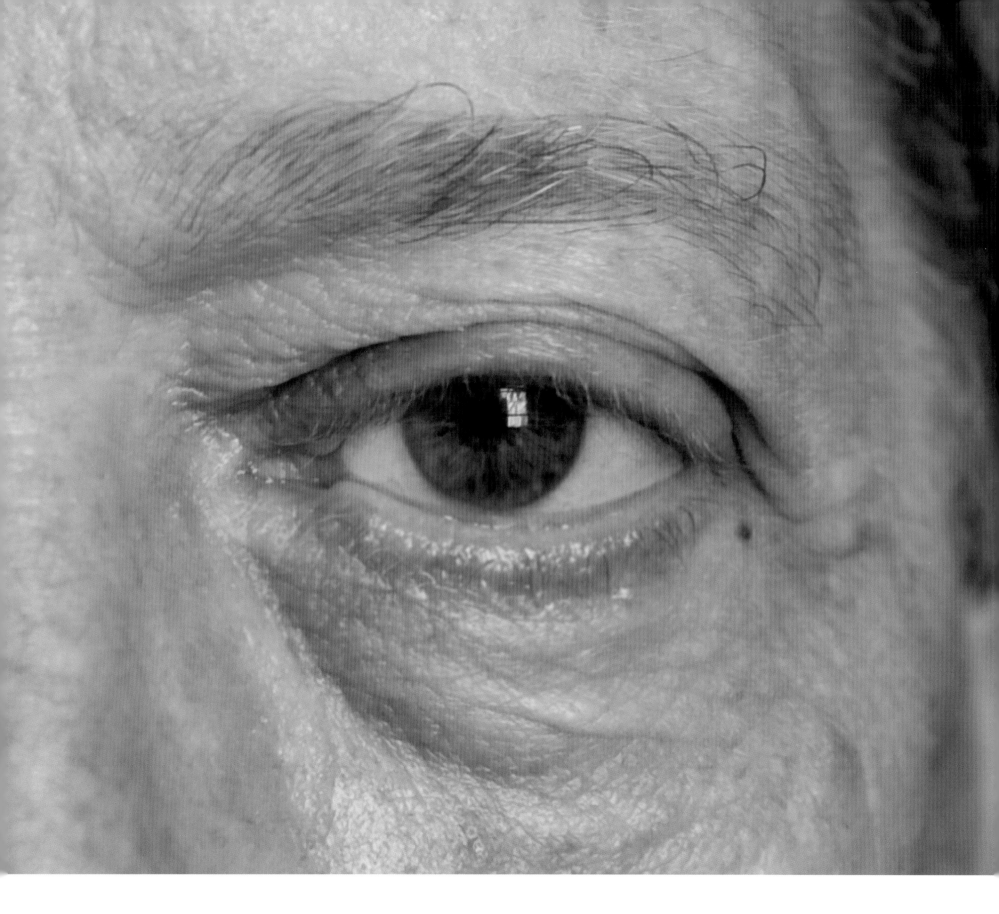

11 | Awareness is everything

Materials define the spaces and forms of architecture. Viewed from a distance, a building may appear as pure shape. At close range, forms and spaces are inseparable from the materials that constitute them. My ambition is not only to create memorable spaces and forms but also to make buildings captivating, physically and emotionally involving their occupants with their surroundings. This involvement with architecture, which draws attention and arouses one's interest in its tactile properties, comes from materials. They supply the corporeal link to conceptual designs. Strong forms and mundane materials or clumsy forms and compelling materials result in just another ordinary structure.

Thinking of a building as merely a shape or a spatial configuration is comparable to judging a book by its cover. Hitching form, space, and materials to each other at the point where their expressive possibilities merge produces remarkable architecture. The rhythms, patterns, textures, and colors of materials embody the limits as well as the spirit of the designer. Regardless of design visions, program requirements, owner ambitions, and the myriad other forces that shape architecture, construction of buildings is always the result of specific material selections. It therefore behooves the architect to accumulate information about as many materials as possible. For me, acquiring knowledge about products and methods of construction is rewarding in itself and has also led to enlightening encounters with manufacturers, artisans, and artists.

Architecture is both an art and a commercial craft; materials bridge the gap between design and economics, the primary forces in construction today. Some products are readily available and others are scarce, but every material can be economically employed.

A material on its own is neither beautiful nor ugly. An unassuming material used in a special way can have considerable visual impact, while an expensive material used in a common way will not show it to full advantage. At the outset of a design process, all materials are ideal—their utilization by architects gives them their visibility and value. As Oscar Wilde so famously put it, "There was no fog in London before Whistler painted it." I have taken great pleasure in using a full range of materials, from standard brick to neglected metals to stone quarry waste. None are out of bounds.

A Zen philosopher once suggested, "We see only what we look at." I enjoy the procedure of direct observation and diligence that helps me judiciously consider new sources and applications of materials. These processes call for intense scrutiny, so intense that it causes the everyday screen of convention to fall away. The world of materials displays itself to those who examine them first-hand, who look at this detail, who explore that product noting their similarities or differences and finding a connection to previous observations. This examination gives the architect the ability to discern the minute but important nuances among similar products.

Awareness is everything. Recognizing the differences among materials is a measured enterprise. Confronting the familiar or the unfamiliar, then noticing the remarkable is the essence of the process. In considering a material, I try to leave everyday reasoning behind and explore every feasible circumstance for its use with a freedom from bias. Sometimes this exploration requires shedding previous assumptions. Le Corbusier wrote, "Reality has nothing in common with books of instruction."

Perception is the beginning of design in architecture. The ability to comprehend the potential of a product is the challenge. My curiosity is always prepared to follow a trail as far as it will go for a material with singular attributes. Customary applications weigh down standard materials; looking beyond the ordinary is crucial. In one architectural generation a given product may be subsumed by uniform application and in another infrequently considered or put to the side. Just as academic design styles may grow stale because of their wholesale use, materials can be diminished and suffer a similar consequence. Over-used items are often malleable enough to become something utterly new, deepening the pool of available resources.

No two people see the same item the same way; even obvious objects are viewed through the lens of one's own visual experience. Moments of perception are often unconnected to anything other than the initial observation.

Opposite: Columbia Public Library, Columbia, MO.; *above:* Cedar Hill Government Center, Cedar Hill, TX.

However, these sporadic insights underlie my personal record of items for future consideration and application. These moments usually flow into one another, forming a collection in a museum of memory where seemingly incompatible items coexist. These stored and sifted perceptions are a plentitude of inspiration and, on a moment's recall, can send my thoughts flowing in an especially productive direction. Andre Malraux wrote, "I keep inside myself, in my private museum, everything I have seen and loved in my life."

Out of a thousand possible perceptions, I seize on only a few to preserve in my memory. These remembrances then circulate and return when new circumstances or a specific set of design conditions arise. Curating and rearranging these notations keeps my original observations alive, and I enjoy placing them inside my mental incubator and watching them evolve. Opportunities for their use can arise at any time. Pursuing new information about materials extends the possibility of drawing unexpected connections among varied elements.

Architecture is always open to appraisal. Structures deemed controversial, aesthetically bankrupt, gross, or humorless when they are built are just as open to further consideration as projects highly regarded from the outset. My interest in assessing older buildings is rooted in a desire to understand the principles underlying a project's design as much as determining its cultural history or chronology. We fully appreciate architecture when we use it or at the very least visit it. The power of example, the chance to learn from another architect's work, is available at every building.

If I can study a building, I consider its architect my contemporary. Examining a structure is almost as good as speaking to the architect, because buildings tell stories through their materials and methods of construction that express the designer's intent. For example, Nicholas Hawksmoor's Saint Mary's Woolnoth, constructed from 1716 to 1724 in the City of London, is as alive with information today as it was when this church first welcomed congregants. Precisely cut stone piled high to form an ascending mass on a triangular site tells the same austere story now as it did when first built, even though the context has drastically altered. We recognize the elemental stone blocks with reveals and incised joints that produce a giant corduroy base supporting a planar transition to a pair of short towers. Hawksmoor's balance between stone blocks and the exaggerated joints between them produces a heavily articulated base enveloping everything in its way: the round headed entry portal, the lunette above, and the attached columns. The contrast between the uppermost twin tower lanterns, four Composite-style columns, and supporting framework couldn't be greater. Connecting rustication and refinement in a compact design continues to pack a powerful visual punch today. This façade reflects the architect's freedom to invent and compose within standard conventions with a readily available material. No architect has replicated Hawksmoor's church, this would be too obvious, but many have simplified and embellished masonry surfaces as he did in this instance. Older structures and their methods of assembly provide insight and inspiration for making new buildings. Going to the place where another architect used a material with exceptional results, one encounters a font of information. Architects remember the past, the obscure, the prominent, the building details, and they recycle them, keeping them alive.

Materials unite two worlds: the world of tradition, the way a product was used in the past, and the world of modernity, the way a material can be used in the 21st century. Today's world is far from stable; materials can be local or come from halfway around the globe. Because most countries share resources, almost any item can become a building product far from its origin.

Right now, in 2008, many materials, such as brick, stone, and wood, have lost their apparent appeal because of their frequent and indiscriminate use. Other construction products, such as glass, resin, and titanium, are identified with

Opposite: Center for Contemporary Arts, Shepherd University, Sheperdstown, WV.
Above and right: Saint Mary's Woolnoth, London, England, designed by Nicholas Hawksmoor.

state-of-the-art design and are popular. In my lexicon of building staples, there is little distinction between the fashionable material and the tired one. Until it is deemed a construction commodity, one material is no better than another.

Despite entire academic programs to the contrary, I find there is little need to invoke literary theories, abstractions, or social circumstances in the selection of a building material. This is not an anti-intellectual bias but rather an invocation of the study of materials as a source of inspiration for architectural design. The products that constitute a building can convey individuality, quality, and innovation.

Using proven technology and ordinary materials to create notable architecture is as constrained an exercise as employing avant-garde techniques and new products. Neither approach is important in itself; the response to the project at hand makes the difference. Prosaic ingredients can look familiar or exotic depending on their usage. The homespun and the radical can exist in the same design, supporting each other as indispensable parts of a whole building. Making ideas bolder, clearer, and stronger assumes complete freedom of selection from a full range of materials.

As the world around us moves toward the uniform and kinetic, architects must keep architecture from becoming anonymous. Architects supply the physical glue that connects form, space, and materials. It is our responsibility to make these choices.

For me, the exploration of all choices is intellectually stimulating and satisfying; it is the path to vivid architecture.

Left: Louise Hopkins Underwood Center for the Arts, Lubbock, TX.

Holzman Moss Architecture, 2007, New York, NY.

Appendix

Project notes

902 Broadway
New York, New York | 1986

This workspace, on two floors of a Flatiron District office building, organizes open workstations, conference rooms, model-making studio, and support facilities along a skewed grid parallel to Broadway, a conscientious departure from the conventional orthogonal floor plan. Circulation routes, defined by colorful, patterned carpets, emphasize the rotation of the plan, in contrast to the original structural columns and exposed ductwork. Residual spaces created by the angular orientation of workspaces provide informal areas for collaboration.

Alaska Center for the Performing Arts
Anchorage, Alaska | 1988

The Alaska Center for the Performing Arts houses three distinct theaters interconnected by multilevel lobbies. Patterned, glass curtain walls animate the facade and provide views to the surrounding Town Square. Twenty-eight regional art and design projects enhance the interior throughout, with each venue—a 2,100-seat multi-use concert hall, an 800-seat drama theater, and a 350-seat theater—boasting its own character. Each theater has its own box office, concession, and support facilities to permit zoning of the Center for either single or multiple productions.

American Film Institute
Kennedy Center
Washington, D.C. | 1973

This project adaptively reuses spaces in the Kennedy Center for a public film theater and the Institute's offices and archives. A variety of formats are screened in the theater; its raised, raked, single-tier seating area offers excellent sightlines. Offices and a small screening room are organized in an L-shaped area using an open plan with freestanding partitions diagonally aligned to differentiate work spaces. Each side of the partitions reveals a different color and lighting treatment to vary the experience of circulating through the space depending upon which direction one walks.

ArtPark
Lewiston, New York | 1974

Artpark was conceived as a temporary, experimental response to housing the visual and performing arts. Its great success, however, prompted the Parks Department to make it a permanent summer feature. This complex, reclaimed from a landfill, consists of an existing performance center and three new elements: "ArtEl," a 500-foot-long elevated artist work space and public promenade; "Town Square," a working environment for large-scale sculpture; and a 300-seat amphitheater overlooking the Niagara River. Much of the project's delight lies in its ability to provide visitors direct access to artists and the creative process.

Best Products Company, Inc., Headquarters, Phase I
Richmond, Virginia | 1980

Governed by a desire to maintain the spirited, non-corporate atmosphere of the owner's former warehouse workspace, the corporate headquarters for Best makes a lively statement in suburban Richmond. The building, which occupies one quadrant of a cloverleaf interstate highway interchange, reflects the curve of its access road. The main facade was designed to expand along the plan's semicircular perimeter, with rectangular private offices jutting out into the landscape at the rear. Inside, the main circulation route is a mosaic-tiled walkway that follows the arc of the façade, providing access to open and closed offices, special spaces, and meeting areas.

Boettcher Concert Hall
Denver, Colorado | 1978

Boettcher Concert Hall is the first American concert hall in which the audience completely surrounds the orchestra. To make this 2,750-seat concert hall intimate and give each member of the audience the best possible visual and acoustic environment, a system of tiered and cantilevered seating envelops the stage. A large-scale, colorful plaid ceiling pattern diminishes the apparent size of the hall and incorporates visible mechanical and electrical systems in its design. Supporting the hall is a modest lobby that provides a variety of overlooks, a rehearsal hall, lounge, dressing rooms, and warm-up area. Boettcher forms part of the four-block Denver Center for the Performing Arts.

Bowdoin College, David Saul Smith Union
Brunswick, Maine | 1995

Conversion of the College's 1912 gymnasium, Hyde Cage, into a student union provides a new and invigorating addition to campus life. Bold patterns, vivid colors, and curvaceous forms make a spirited statement, adding elements of surprise and vigor within the historic building's staid envelope. Careful modulation of size and scale allows the building to comfortably accommodate solitary activities, small group meetings, and campus-wide social events. From vantage points along its balconies, the building permits glimpses of the activities within and the campus beyond.

Brooklyn Children's Museum
Brooklyn, New York | 1977

The Brooklyn Children's Museum is submerged below grade (with one exposed corner) to allow the adjoining Brower Park to extend across its roof. A playground and open-air theater are also part of its exterior. Inside, an open-plan participatory learning environment with four exhibition levels offers diverse, flexible spaces where children have hands-on engagement with objects in the museum's collection. Workshops, resource library, dance studio, photography studio and darkroom, and marketplace provide additional learning opportunities.

Cleveland Public Library, Louis Stokes Wing
Cleveland, Ohio | 1998

Components of the three-phase project included the Louis Stokes Wing, renovation of the existing 1926 Main Library, and reconstruction of the Eastman Reading Garden. The 10-story Stokes Wing is constructed of stone and glass and complements the other civic buildings in the historic 1903 Group Plan. It enlarges the Main Library by some 255,000 square feet, nearly doubling its size and accommodates additional library materials, state-of-the-art technology, and a 250-seat auditorium. The two buildings connect underground and are separated at grade by the enhanced Eastman Reading Garden, an intimate and reanimated beloved public park.

Columbia Public Library
Columbia, Missouri | 2002

The intent of the design was to make the library a focal point of the downtown community and to create spaces that were highly functional as well as comfortable and enjoyable. The design integrates the structure of the earlier 1971 library into four distinct geometric forms that house specific library functions: a conical entrance pavilion, an oblong utility spine, a semi-cylindrical collections area, and an orthogonal administration wing. Each is clad in a different material to signify its individual function and to add to the entire composition. The striking, identifiable forms and creative use of materials produce strong impressions for both users and staff.

Courtyard Theater
Plano, Texas | 2002

The Courtyard Theater is a new 325-seat venue that provides an environment for performances, lectures, and events by existing local organizations as well as touring groups. The theater is housed in the historic Cox High School gymnasium annex, which was constructed in 1935 under the Works Progress Administration program. Inside, the gymnasium was transformed into a flexible performance space that can be arranged to allow frontal, thrust, or theater-in-the-round style presentations. This high level of adaptability is a result of design elements such as large movable acoustical panels that help tune the room for each specific function.

Creighton University, Lied Education Center for the Arts
Omaha, Nebraska | 1996

Creighton University's fine and performing arts departments, previously scattered across the campus in adapted spaces, are consolidated in the Lied Education Center for the Arts. In its plan and forms the building symbolizes the diversity and vitality of these activities. The Center is a stepped cube of stone, buff brick, exposed concrete, and glass curtain wall. Three floor levels, terraced into the sloping site, house an art gallery and flexible performance and teaching studios, labs, and offices. A 350-seat proscenium theater is at the physical and programmatic heart of the building.

Firemen's Training Center
Ward's Island, New York | 1975

The Firemen's Training Center serves as an academic and physical training facility for all New York City firemen. The complex includes an education/administration building and eight training/service facilities on landfill between Ward's and Randalls Island. Because of their function and requirement for low maintenance training and service buildings—replicas of a loft building, a tenement, and a frame dwelling—are simple concrete structures faced with brick. The metal-clad education/administration building houses classrooms, administration areas, a library, lunchroom, firemen's store, and orientation and exhibition areas.

Fox Theatres
Wyomissing, Pennsylvania | 1993

Poised at the curve of an elliptical drive on an 8.5-acre site, Fox Theatres sits on a black-and-white striped plaza made of alternating bands of concrete and asphalt. Eight theaters are grouped into a simple L-shaped volume enclosed by massive, corrugated galvanized siding. Against this backdrop an upturned roof shelters a glass-walled lobby intersected by a towering box-office that serves as a beacon to all movie-goers. The dynamic interplay of lights, colors, patterns, and forms activates the theater complex and heightens the cinematic experience.

George A. Purefoy Municipal Center
Frisco, Texas | 2006

The George A. Purefoy Municipal Center is the centerpiece of the 145-acre Frisco Square, a master planned extension of the downtown area. To endow the building with civic presence, the complex was designed with a strong stone facade and striking central clock tower that houses a public City Room at its summit. The City Hall includes a 315-seat Council Chamber and city offices. Features of the new Public Library include spaces for collections, reference area, and Family Learning Center, in addition to various administrative offices, conference rooms, a café, and a gallery.

Globe-News Center for the Performing Arts
Amarillo, Texas | 2006

The Globe-News Center for the Performing Arts was borne out of the community of Amarillo's desire to have a state-of-the-art auditorium that that could house orchestral and theatrical performances, showcasing both functions with equal ease and excellence. At its heart is the 1,300-seat Carol Bush Emeny Performance Hall; its innovative, one-of-a-kind design creates optimal acoustical environments for various types of performances. Supporting the auditorium are education, rehearsal, and back-of-house spaces. The Center's undulating roofline reflects the nearby majestic Palo Duro Canyon, establishing itself as regional landmark in the local landscape.

Hawaii Theatre Center
Honolulu, Hawaii | 1996

Proclaimed the "Pride of the Pacific" when it opened in 1922, restoration of the Hawaii Theatre was considered pivotal to the renaissance of Oahu's historic Chinatown district, adjacent to downtown. New sound, lighting, and projection systems were carefully integrated, and its proscenium mural was restored and recreated, as were its many Shakespearean-motif trompe l'oeil images and bas-reliefs. To meet accessibility requirements and to improve sightlines, a parterre was added along with new seats. To bring the rest of the 1,400-seat theater up to date, the wing space and the orchestra pit were enlarged with a hydraulic lift, and a large function room, concession stands, and dressing rooms were added. A subsequent phase of work returned the historic marquee and a final phase will add backstage and support space.

Highland House
Madison, Wisconsin | 1993

Highland House comprises seven polygonal structures, connected by arc-shaped "meridians" that allude to Italian hilltop villas. Each element of this small village was carefully sited and sculpted to relate to its landscape, protecting neighbors' views and preserving the topography. The connecting meridians are covered with red metal shingles that set them apart from the other structures, which are sheathed in granite, limestone, and wood clapboard. Towers and vivid rooflines define the massing and are fenestrated to permit generous natural interior light and panoramas of earth, sky, and in winter, Lake Mendota.

Holman Kitchen
New York, New York | 2000

Redesign of this existing kitchen, a 6.75-by-12-foot sea of white laminate and oversized, floor-to-ceiling cabinets, creatively provides new preparation, storage, and dining space. A main element is the curving, overlapping custom cabinetry, which gives the illusion of space by drawing the eye around the room and adding visual interest. Ebonized walnut, chosen for its warmth, extends to the countertops, rather than colder materials like granite, marble, or stainless steel. The industrial rubber flooring is both durable and easy to clean, but moreover offers rich texture and color—green with bands of yellow in a soft matte finish.

Hult Center for the Performing Arts
Eugene, Oregon | 1982

Design for the Hult Center for the Performing Arts includes the 2,500-seat Silva Concert Hall, 500-seat Soreng Theater, and performance support spaces. Both venues are linked by a four-level lobby defined by a series of peaked, skylit gables supported by massive timber columns and beams. Viewed from the outside, the lobby's dark and jagged silhouette conjures up the nearby landscape. As seen in the evening, the interior illumination is revealed in stripes, varied by the color and texture of the bands of glass. It serves as its own marquee, lighting up the streets and giving the City of Eugene a lively urban identity.

ImaginOn: The Joe & Joan Martin Center
Charlotte, North Carolina | 2005

ImaginOn is a unique prototype for education that allows young people to engage with the written, spoken, and electronic word. Fusing library and theater programs into an integrated learning environment, the facility houses 570- and 250-seat auditoriums, children's and teen's library spaces, a blue-screen animation theater, a children's storytelling room, and an interactive exhibit. Unusual spatial configurations result from interaction between geometric elements—a parallelogram, a helix, cubes, and a rotated ellipse. Each element is clad in a unique material: glass, metal, stone, concrete, and recycled clay tiles, creating a stimulating environment rich in textures, colors, and patterns.

Landsman/Holzman Loft
New York, New York | 1986

In transforming a century-old industrial loft into a residence, care was given to preserve remnants of the past, while creating a highly personal, idiosyncratic living space. A number of elements were left as they were, including the distressed beams and ceiling joists that show previous installations for a hoist and crane rail, and scarred wood flooring now sanded and stained. This conservation of historic features contrasts with prevalent and unusual additions. Stained flakeboard installed with grommets, corrugated fiberglass mounted on tiny wire joists, and galvanized metal sheets with brick and stone patterns are just some of the room enclosures that are animated by changing natural light and a variety of artificial light.

Los Angeles Public Library: Central Library
Los Angeles, California | 1993

When the Central Library opened in 1926, it was the most significant civic building in Los Angeles. Its design, by Bertram Goodhue, was progressive in spirit, without specific relation to any defined historical style. Renovation restored artworks, recaptured original colors, and re-created historic light fixtures. In designing the 330,000-square-foot Tom Bradley Wing, the landmark's prominence in the skyline was retained by placing more than half of the new addition below street level. The central feature of the new wing is a dramatic, glass-roofed, eight-story atrium, a modern counterpart to the original central rotunda. Outside, the new Robert F. Maguire III Gardens top a 942-space underground parking garage.

Los Angeles County Museum of Art, Robert O. Anderson Building
Los Angeles, California | 1986

The Robert O. Anderson Building addition to the Los Angeles County Museum of Art provides exhibition galleries for the museum's 20th-century art collections and special exhibitions as well as support and administrative office space. The building redefines the museum's presence on Wilshire Boulevard and establishes a single identifiable entrance to this major civic complex. Its facade of stone, glass block, and terracotta relates directly to Wilshire's Art Deco and Art Moderne heritage. The three-story Times Mirror Central Court, covering nearly an acre, visually and symbolically unites the four buildings that make up the museum.

Middlebury College, Center for the Arts
Middlebury, Vermont | 1992

The 100,000-square-foot Center for the Arts provides a single focal point for the arts, accessible to both the Middlebury community and the general public. In consolidating the arts programs within one facility, care was taken to give each department its own identity while providing common social spaces for artistic cross-fertilization. The building is inset and overlaid with square, circular, and octagonal forms that represent each of the major program elements—concert hall, studio theater, dance studio, and art gallery—and protrude out from a common skin and a distinct, standing seam copper roof.

Occupational Health Center
Columbus, Indiana | 1974

This health center serves Cummins Engine with a design based on a configuration of open, semi-open, and closed spaces defined by two overlapping grids. Closed, private activities are set around the building's perimeter, sheathed in black reflecting glass. Open, public activities are set within the internal grid, whose edges delineate separate public and staff circulation; these paths are outlined by a continuous skylight in mirrored glass. Brightly colored mechanical and structural systems accentuate the building's various layers.

Ohio Theatre and Galbreath Arts Pavilion
Columbus, Ohio | 1984

Now completely restored, the Ohio Theater, built by Thomas W. Lamb in 1928, is recognized as the official state theater of Ohio. Transformation of this historic 3,000-seat movie theater into a full-fledged performing arts center resulted from several modifications and expansions that enlarged the stage and added a new loading dock, lobby, and arts pavilion to house dressing rooms, support spaces, and rehearsal facilities, as well as administration space for local arts organizations. The six-story pavilion, designed in an architectural vocabulary compatible with Lamb's original architectural fantasy, can be used independently of the theater for receptions and community events.

Orchestra Hall
Minneapolis, Minnesota | 1974

The design directive for this 2,700-seat hall was an environment of superior sound and design, which would enhance the experience of both attending and performing in concerts. The hall itself and the support spaces use separate architectural vocabularies, set within different geometric grids. This organization articulates the building's functions and spaces, and expresses them as individual volumes. The red-brick hall is set back at a slight angle from the street to create a new landscaped plaza, while the steel-and-glass structure housing the lobby contains a series of catwalks and platforms from which audiences are offered views of the city.

Phillips Exeter Academy, Fisher Theater
Exeter, New Hampshire | 1972

Designed for various types of student productions, this 260-seat theater invites participation with relaxed divisions between performers and audience, simple scenic devices, and easy access to lighting equipment and the stage area. Five regular structural bays of a standard prefabricated building for the theater are integrated in an offset configuration, breaking up the system's rectangular volume and creating spaces that are diversely shaped, both inside and out. Support spaces include a shop, costume fabrication room, makeup and changing spaces, rehearsal areas, offices, conference rooms, and classrooms.

Punahou School, Dillingham Hall
Honolulu, Hawaii | 1994

Designed by Bertram Goodhue in 1924, and completed after his death by Hardie Philip in 1929, Dillingham Hall has been in continual use since the day it opened. As part of the restoration, layers of contemporary systems were used as ornament to essentially complete the room with respect to Goodhue's legacy for rich detail and surface decoration while upgrading the theater to contemporary standards. A vigorous palette of tropical green and blue was introduced in patterned upholstery, textured carpeting, a custom-designed stage curtain, and new technical arches and mechanical stanchions. Seating for 653 patrons was reconfigured to improve sightlines. A new annex adds faculty offices and shop spaces and increases workspace and storage.

Ramapo College of New Jersey, Salameno Spiritual Center
Mahwah, New Jersey | 2007

This modest spiritual center, situated adjacent to a pond in the heart of campus, provides the college community with a place for reflection, meditation, celebration, and assembly. At its core is a flexible room accommodating 80 people. Places for quiet meditation further lend an oasis of calm, offering respite from the pressures of daily life, and define by their placement a central outdoor space.

Ramapo College of New Jersey, The Angelica and Russ Berrie Center for Performing and Visual Arts
Mahwah, New Jersey | 1999

The Berrie Center features spaces for theater, video art, electronic music, computer design, digital imaging, virtual reality, and a range of multi-media applications. These are contained in four geometric volumes, each distinguished by its own roofline and materials. A long, low rectangular wing houses galleries, classrooms, practice rooms, studios, and shops as well as administration and faculty offices. The other three building forms—a proscenium theater, a rehearsal/performance space, and a studio theater—intersect it. Their unique volumetric shapes identify them from the exterior as significant program spaces for experimental art.

Salisbury Upper School
Salisbury, Maryland | 1997

In 1972, the firm designed the Salisbury Lower School, a progressive learning institution for Pre-K to 8th-grade students. Its success led to the desire, 25 years later, for a new upper school. Several similarities exist between the two generations of buildings. Interior plans in both feature partitions, set on rotated axes, that define open and closed areas for a variety of teaching situations, and amorphous residual spaces for less structured encounters. Brightly colored beams, columns, and railings, vigorously patterned carpets, and skylights intensify each learning environment. At the center of the upper school is a domed structure that encloses the library, administrative offices, and informal gathering spaces. From this "hub" radiate a vehicular drop-off area, an academic wing, an athletic wing, and space for a future expansion.

San Angelo Museum of Fine Arts and Education Center
San Angelo, Texas | 1999

SAMFA serves as a center of culture, education, and entertainment in its community. It is jointly used by Angelo State University, whose art department conducts regular studio art classes at the museum. Facilities include galleries, an education center, an outdoor kiln yard, a 300-seat public meeting space, a gift shop, a library, a multi-level lobby, and administrative offices. In its elongated shape, materials, color, and texture, the museum resembles the limestone buildings at nearby Fort Concho. A spavine-vaulted, copper-clad roof distinguishes the building from a distance, allowing it to become another of the city's landmarks.

Shaw University, Master Plan and Community Services Center
Raleigh, North Carolina | 1970

The master plan juxtaposes new and old construction in a geometric manner to invigorate and diversify the campus. The Community Services Center, the first phase of construction, utilizes pre-engineered buildings as an early demonstration of the speed, economy, and flexibility of ready-made building enclosures. For use by all college departments, the center provides space for seminars, workshops, community gatherings, theater, and individual student activities. Two units overlap on the diagonal to establish an irregular configuration inside and out, and fracture the system's rectangular appearance.

Shepherd University, Center for Contemporary Arts
Shepherdstown, West Virginia | 2008 Phase I

The new Contemporary Arts Center is an innovative joint initiative between Shepherd University and the Contemporary American Theater Festival (CATF), to explore and showcase contemporary art and new theater in Shepherdstown. The Center is composed of multiple buildings that house two 250-seat theaters, a 150-seat flexible rehearsal and performance space, an art gallery, fine arts studios, labs, and classrooms. Its design—agrarian barn profiles above large steel trusses with exterior copper sheeting—strikes a balance between the urban and rural, and the academic and creative. Dog trots, screened porches, and open terraces enable use as student venues or highly public facilities for events of all forms.

Skyler House
Bridgehampton, NY | 1985

This house is a composite of vertical masses with distinct profiles on all four sides and varying geometric fenestration throughout. The most open areas front the beach and ocean, reflecting the home's orientation. The interior is staggered to create four distinct half-levels, a plan that separates the living and eating areas on an elevated first floor and provides the master bedroom with greater privacy. The interior is expressed on the exterior by an asymmetrical composition of differently sized windows. This window pattern creates a variety of natural light conditions and controlled views from room to room.

Temple Israel
Dayton, Ohio | 1995

Situated in a picturesque, green space along the Great Miami River adjacent to downtown, the synagogue commands a spectacular view of the skyline. Temple Israel houses a sanctuary, a great hall, a chapel, classrooms, a library, a gift shop, a kitchen, and other educational and administrative spaces. The convergence of rooms, each with a distinct geometry, results in a series of connected chambers and passageways that function as an active lobby space. Within the plan of closely related rooms, the sanctuary is its fulcrum. Adjacent to the sanctuary, the east wall and roof of the great hall follow a cosine curve; from the exterior, the undulating wall responds to the gentle flow and bends of the river it faces.

Texas A&M University-Corpus Christi, Performing Arts Center
Corpus Christi, Texas | 2005 Phase I

The Performing Arts Center reflects the University's commitment to enhancing the cultural environment on campus, cultivating new curricula in the performing arts and providing exceptional cultural facilities for Corpus Christi and the larger South Texas community. The building takes advantage of its scenic location along Corpus Christi Bay by offering incomparable views from the multi-storied public areas. The 1,500-seat concert hall is conceived as a teaching facility, though it is also used by the local symphony. A three-ringed seating configuration promotes intimacy at times when the room is not at full capacity.

Texas Christian University, Mary D. and F. Howard Walsh Center for Performing Arts
Fort Worth, Texas | 1998

The performance spaces for the music and theater department include a 350-seat recital hall and a 200-seat studio theater along with complementary rehearsal rooms, teaching studios, practice rooms, music library, shop, and support spaces. A multi-level lobby provides access to all new public spaces and the adjoining Landreth Hall. The recital hall closely reflects shapes and materials of other buildings along the formal public lawn that it fronts, while the studio theater, which faces an interior residential quad, features asymmetrical massing and prominent materials. These volumes are contained within the overall building mass of grooved, yellow brick, tying together generations of campus architecture in a contemporary form.

Texas Tech University, Student Union Building
Lubbock, Texas | 2006

A new series of pavilions offer spaces for dining, socializing, studying, and group meeting, responding to the significant enrollment growth and changes in student life on campus. The interior environment of the pavilions and the renovation of the existing structure provides many new and enlarged spaces for student government, organizations, retail, and special uses, but most importantly, the building now has gathering points for social interaction, an easily understood vertical and horizontal circulation path through the entire structure, and a sense of natural light not present before. For the first time in its history, there are views out to some of the most memorable surrounding buildings on campus, and views of the life and activities within.

University of Nebraska at Omaha, Del and Lou Ann Weber Fine Arts Building
Omaha, Nebraska | 1992

The new Fine Arts Building houses the departments of Theater, Art, and Creative Writing. Teaching, lab, and office spaces are arranged in a linear fashion, forming a spine along which varied geometric elements are appended. Hexagonal towers house faculty and student lounges and provide a circulation hinge to which a future addition is possible. The trapezoidal experimental theater and rectangular gallery wing, which veer off the building's west side, shape a new sculpture garden. A multi-story lobby to the north completes the 78,500-square-foot composition.

University of North Texas, Lucille Murchison Performing Arts Center
Denton, Texas | 1999

The 72,500-square-foot Murchison Center accommodates student concerts, performances by guest artists, musical theater and opera productions, and special events. It houses a 1,100-seat concert hall, a 400-seat lyric theater, rehearsal rooms, a music library, a recording studio, dressing rooms, green room, and administrative offices. The center's location, on the western edge of the campus along Interstate 35, is highly visible. As a result, the building becomes a gateway to the University, formally welcoming visitors. Its unique forms and faceted zinc-clad dome make it a recognizable destination as cars approach.

University of the South, McClurg Hall
Sewanee, Tennessee | 2001

McClurg Hall was designed to serve all undergraduate dining needs with a formal 450-seat refectory that reinforces the campus tradition of dining together, and a 250-seat informal dining room that allows for smaller, more intimate groups to gather. These are supported by a servery, a state-of-the-art kitchen, meeting rooms, and administrative spaces. The building is a distinct and modern addition to the campus, while honoring its Gothic spirit with an emphasis on the vertical and through the use of stone and natural light.

University of Toledo, Performing Arts Center
Toledo, Ohio | 1967

The building's site is a former parking lot, and its central location at the intersection of student and public functions helps unify the campus. Sloping roofs and landscaping visually link the center with the disparate collection of buildings that surround the site. Its two major elements are a 500-seat theater and a 500-seat concert hall, both designed with unusual shapes that promote a sense of intimacy. The theater's eight distinct seating fragments focus on a thrust-stage performance platform. In the concert hall, half the seating is in boxes surrounding the orchestra level.

Virginia Museum of Fine Arts, West Wing
Richmond, Virginia | 1985

This distinct fourth addition to the 1936 museum building adds 90,000 square feet of exhibition and support space while blending with the plan of the original. The museum's original axial plan is referred to in the new West Wing's central hall. However, the addition's plan then diverges with a series of small-scale traditional spaces and large-scale galleries that complement the character of two vastly different collections of paintings and decorative arts. The Wing's smooth and ribbed exterior limestone textures arranged in bands advance or recede to heighten shadow and depth, a composition that complements the original limestone and brick neo-Georgian building.

WCCO–TV Communication Center and Headquarters
Minneapolis, Minnesota | 1983

Small in size, but forceful in demeanor, the WCCO building has a great presence, even a sense of monumentality. A three-story stepped structure, the building serves as a pedestal for the constantly changing panoply of rooftop communication technology that announces its function. Satellite dishes, microwave screens, electronic interference shields, and weather equipment are mounted on the roof. This contrast between the solid base and high-tech equipment is one of the structure's most distinctive features. The facility includes a newsroom, a production studio, editing suites, control facilities, offices, and underground parking.

Willard Hotel, Office, and Retail Complex
Washington, D.C. | 1984

Expansion and restoration of the 1904 Willard Hotel extends its traditions into a new century. An addition composed of four descending segments echoes the cornices, pediments, and corners of the original French Second Empire-inspired landmark in massing, scale, and details to create a cohesive overall design. These segments step up and back as they approach the original building to create an outdoor plaza. The completed project includes a 400-room luxury hotel, 210,000 square feet of office space, 40,000 square feet of stores and shops, and new underground parking.

Collaborators

Holzman Moss Architecture

Fermin Beltran

Steve Benesh

Jessica Blum

Ben Caldwell

Patty Chen

Michael Connolly

Darwin Harrison

Malcolm Holzman

Mathew Kirschner

Eddie Kung

Ching-Wen Lin

Brad Lukanic

Douglas Moss

Chiun Ng

Jose Reyes

Margaret Sullivan

Debbi Waters

Hardy Holzman Pfeiffer Associates

Yasin Abdullah	Diane Blum	Neil Dixon
Cleveland Adams	William Bolling	Violeta Dumlao
Ron Albinson	Nestor Bottino	Milton Ewell
Dorothy Alexander	James Brogan	Harris Feinn
Robert Almodovar	Paul Buck	Jaime Fournier
Christopher Bach	Ryan Bussard	Darlene Fridstein
Shane Baker	Evan Carzis	Jean Gath
Curtis Bales	Mario Censullo	Nancy Geng
Daniel Barrenechea	Hakee Chang	Alec Gibson
Kim Beeler	Jonathan Cohn	Robert Goesling
Caroline Bertrand	Mark DeMarta	Victor Gong
Donald Billinkoff	James Despirito	Theron Grinage

David Gross
Robert Gross
Perry Hall
Victoria Hammer
Hugh Hardy
John Harris
Lee Harris
M. Hernert
Randolph Hicks
David Hoggatt
Matt Jasmin
Matt Jogan
Peter Johnson
Stephen Johnson
Stewart Jones
William Jordan
Chris Kaiser
Carl Karas
Mark Kessler
Patricia Knobloch
Kurt Kucsma
Paul Kulig
Robin Kunz
Tony LaFazia
Diane Lam
Don Lasker
Jeeyoon Lim
Dan Lincoln
Pamela Loeffelman

Rob Lopez
Joyce Louie
Hilda Lowenberg
Raoul Lowenberg
John Lowery
Jeff Lynch
John Maddox
Leah Madrid
Jamie Margulies
John Mariani
Jack Martin
Todd Martin
Mike McGlone
Manuel Mergal
Ky Mikagi
Catherine Minervini
Charles Muse
Jeff Neaves
Kristopher Nikolich
Mindy No
Setrak Ohannessian
Ryoko Okada
Susan Olroyd
Scott Perry
Norman Pfeiffer
Daria Pizzetta
Susan Pon
Jeff Poorten
Lynne Redding

Lindsey Reeds
Candace Renfro
James Rhodes
Allen Robinson
Victor Rodriguez
Candace Rosean
Michael Ross
Gilbert Sanchez
James Sarfaty
Maya Schali
Kala Somvanshi
Bruce Spenadel
Steven Stainbrook
Patrick Stanigar
Jonathan Strauss
Mark Tannin
Matt Tendler
Dale Turner
Binita Vijayvergiya
Amy Wagerbach
Kristina Walker
John Way
Marvin Wiehe
Peter Wilson
Amy Wolk
Winslow Wu
Robert Yorke

Photography credits

8–9 *Stardust* sheet music by Hoagy Carmichael; sheet music from the collection of Malcolm Holzman
11 © 2006, Hester + Hardaway
12 © Christopher Arend
13 © Christopher Little
14 © Malcolm Holzman
15 © Holzman Moss Architecture
16 © Tom Kessler
17 © Holzman Moss Architecture
18–19 © Jose Fuste Raga/Corbis
20–21 © HHPA
22 © Tom Kessler
23 © Craig Blackmon
24 right Photo by Hedrich-Blessing, courtesy Chicago Historical Society/Chicago History Museum
24 center Postcard from the collection of Malcolm Holzman
24 left Photo from the Toledo-Lucas County Public Library, *Images in Time Collection,* used with kind permission
25 © HHPA
26 top © HHPA
26 bottom © Richard Rose
27 Bruce A. Goff Archive, Ryerson and Burnham Archives; reproduction © The Art Institute of Chicago
28–29 © Tom Kessler
30, 31 Catalogue courtesy Gladding McBean & Co.
32–33 © Cervin Robinson
34 © Cervin Robinson
36–37 © Tom Kessler
38 © Malcolm Holzman
39 left Photo courtesy Gladding McBean & Co.
39 right © Meg Carlough
40 © Elliott Kaufman
41 © HHPA
43 © Ngoc Minh Ngo
44 left Photo by Abbie Rowe, National Park Service/John Fitzgerald Kennedy Library, Boston
44 right © Carol Highsmith
45 © 1986, David Franzen
46 © 2006, David Franzen
47 © Tom Kessler
48 top © Meg Carlough
48 right © Gary R. McCauley, Cincinnati, OH

49 © Cervin Robinson
50–51 © Peter Brenner
52 © Jessica Blum
53 *Niagara,* 1857, Frederic Edwin Church, 42½" x 90½", oil on canvas; courtesy Corcoran Gallery of Art, Washington D.C., accession no. 76.15, Museum Purchase, Gallery Fund
54 top © Nicholas Whitman, www.nwphoto.com
54 bottom © Balthazar Korab Ltd
55, 56–57 © Norman McGrath
58 © Tim Street-Porter
59 © Norman McGrath
60–61 © Tom Kessler
62 © Norman McGrath
63 top © Norman McGrath
63 bottom © Malcolm Holzman
64 © Craig Blackmon
65 © Craig Blackmon
66–67 © Craig Blackmon
68–69 © Norman McGrath
70 © Cervin Robinson
71 © Patricia Layman Bazelon
72 © Cervin Robinson
73 © HHPA
74–75 © 2006, Hester + Hardaway
76–77 © Tom Kessler
78 Courtesy Revere Copper Products, Inc.
79 top © HHPA
79 bottom © Craig Blackmon
80 © Norman McGrath
81 © Malcolm Holzman
82 top © Meg Carlough
82 bottom © Norman McGrath
83 © Christopher Little
84 © Malcolm Holzman
85 © Tom Kessler
86 left © Tom Kessler
86 right © Malcolm Holzman
87 © Norman McGrath
88 right © Brian Vanden Brink
88 left © Tom Kessler
89 © Tom Kessler
90 © HHPA
91 © Craig Blackmon
92–93 © 1998, David Franzen
94 © Foaad Farah
95 top © 1998, David Franzen
95 bottom © Tom Kessler
96–97 © Christopher Lovi

98–99 © Norman McGrath
100 © Malcolm Holzman
101 © Malcolm Holzman
102 © HHPA
103 © Tom Kessler
104 © Tom Kessler
105 left © HHPA
105 right © Norman McGrath
106 © Norman McGrath
107 © HHPA
108 © Norman McGrath
109 © 1998, David Franzen
110 © Craig Blackmon
111 © Tom Kessler
112 © HHPA
112–113 © Tom Kessler
115 © Norman McGrath
116–117 © 2006, Hester + Hardaway
119 © Malcolm Holzman
120 © HHPA
121 © Tom Kessler
122–123 © Tom Kessler
124 © 1998, David Franzen
125 © 2006, Hester + Hardaway
126–127 © 1998, David Franzen
128 © Tom Kessler
129 © Malcolm Holzman
130–131 © Tim Hursley
132 © Malcolm Holzman
133 © Tom Kessler
134–135 © Tom Kessler
136–137 © Tom Kessler
138 © 2006 Artists Rights Society (ARS), New York/ADAGP, Paris/Succession Marcel Duchamp, Marcel Duchamp, *Fountain,* 1917, Photographed by Alfred Stieglitz, Philadelphia Museum of Art: The Louise and Walter Arensberg Collection, 1950, 1950-134-92, detail
138–139 Cover artwork © Sony BMG Music Entertainment, used with kind permission
139 *Half Face with Collar,* 1963, Roy Lichtenstein, oil and magna on canvas, 48" x 48", 121.9 x 121.9 cm, © Estate of Roy Lichtenstein; Photo: Robert McKeever
140 left Postcard from the collection of Malcolm Holzman
140 right © Malcolm Holzman